Blood relations

Blood relations

BLOOD GROUPS AND ANTHROPOLOGY

A. E. Mourant

MA, D. Phil., DM, FRCP, FRC. Path., FRS
Formerly Director, Serological Population Genetics Laboratory, St Bartholomew's Hospital, London

OXFORD UNIVERSITY PRESS

Oxford University Press, Walton Street, Oxford OX2 6DP
London New York Toronto
Delhi Bombay Calcutta Madras Karachi
Kuala Lumpur Singapore Hong Kong Tokyo
Nairobi Dar es Salaam Cape Town
Melbourne Auckland
and associated companies in
Beirut Berlin Ibadan Mexico City Nicosia

OXFORD is a trade mark of Oxford University Press

Published in the United States
by Oxford University Press, New York

First published 1983
First issued in paperback 1985

British Library Cataloguing in Publication Data
Mourant, A. E.
Blood relations.
1. Blood groups
I Title
612'.11825 QP98
ISBN 0-19-857631-5

Printed in Great Britain by
Thetford Press Limited, Thetford, Norfolk

Preface

Until the time of the Second World War, physical anthropology meant the comparative study of the anatomical characteristics of the body in different human populations and individuals. It signified particularly the measurement of the body and its parts, especially the skull, in living and dead persons.

Since then, however, arising largely from the intensive development of blood transfusion promoted by the war, the blood groups, supplemented more recently by other hereditary biochemical characteristics, have come to provide an alternative means of classification of the living which has to a great extent replaced direct body observations.

As a scientific tool, the blood groups, with their precisely understood genetics, are probably easier to use than physical measurements, but the reasons behind their application are far less obvious to the layman.

The objects of this book are to explain, in as simple terms as possible, how blood-group anthropology works and how it is applied to particular populations, and to set out some of the conclusions that can be drawn from it as to relations between populations.

In a final chapter the wider implications of the subject are examined, such as relations between genetic characteristics, susceptibility to particular diseases, and exposure to various environmental factors.

Every effort has been made to ensure that the statements and conclusions of this book are as accurate as possible. Readers are assured that this will not later have anything to unlearn, but they are not asked to take this on trust. A full bibliography would run to many thousand references, but a short select bibliography has been included. Its primary purpose is to serve as a reading list, though occasional references are made to it in the text. Readers who wish for fuller documentation should consult the author's books, *The distribution of the human blood groups, Blood groups and diseases*, and *The genetics of the Jews*, listed in that bibliography.

The present book is based, to a considerable extent, upon the works just mentioned, written in collaboration with Dr Ada C. Kopeć and Mrs Kazimiera Domaniewska-Sobczak, to whom I express my warmest thanks. They are, of course, in no way responsible for the present work. I am most grateful also to Mr G. Misson and Professor D. F. Roberts, and to the Staff of the Oxford University Press, for reading the whole text and

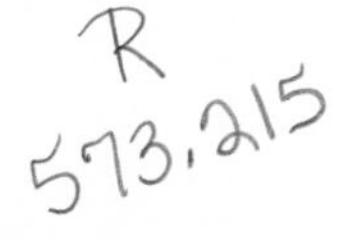

making constructive suggestions, most of which I have adopted. I am also deeply grateful to my wife for typing several drafts of the book from my untidy manuscripts. I am indebted to Mrs. Marie Crookston for suggesting the title of the book.

Jersey A. E. M.
March, 1982

Contents

1. How we recognize people 1
2. Elementary genetics 4
3. Africa 36
4. Asia 59
5. Europe 73
6. The Jews and the Gypsies 90
7. The Pacific islanders 99
8. Amerindians and Eskimos 108
9. Causes of gene frequency change 115

Further reading 131

Maps 133

Index 141

1. How we recognize people

Inherited and acquired characteristics

One of the most important faculties of human beings, as indeed of all 'higher' animals, is that of recognizing individual members of their own species. It is a faculty which we all take very much for granted, and if one of us were unexpectedly asked how we did it, even the immediate and perhaps superficial answer would require some moments' thought.

Facial conformation is, of course, all important, yet notoriously difficult to describe, especially when it is a matter of distinguishing between two or more members of one's own population or between persons of the same 'race'. One reason for this is that the mind, in interpreting the results of visual observation, pays considerable attention to 'expression' which consists of subtle indications of the emotions of the person observed, brought about by slight contractions of the facial muscles. We have all seen a person alter in appearance almost out of recognition by the sudden removal of some source of worry, such as an unhappy personal relationship, yet the main shape of the face, and certainly the shape of the bones which form its main structure, have changed not at all. It is for this reason that the interpretations of the sensitive painter convey much more to most of us than the scientific descriptions of the anthropologist.

Yet for scientific purposes we need to be able to describe the individual in unambiguous terms which will mean the same to all readers.

We have so far mentioned only the shape of the face, because it is at the same time the most important characteristic in everyday life, and also perhaps the most difficult to describe. But the shape and size of the body as a whole, and of its parts, are also of obvious descriptive importance. Colour of eyes and hair and, when we come to major 'racial' distinctions, colour of skin, are also of outstanding importance, as is the texture of the hair. These differences in colour and texture are much more easy to put into words than are subtle anatomical distinctions. The former, together with obvious 'racial' anatomical characteristics such as slanting eye openings and everted lips, form the basis in most people's minds of a separate set of distinctions of a quite different kind from those which we use in distinguishing individuals of our own race. Later on we shall consider the question of whether these really constitute two separate scientific levels of description and distinction.

The characteristics just described are mainly inborn and hereditary rather than acquired, but some, especially body weight and the contours of the soft parts of the body, are mainly acquired, and more dependent upon the state of nutrition. Skull shape can be modified permanently by pressures applied during infancy.

To change from morphological and visual characteristics to auditory ones, language and accent are learned, but the characteristic sound of the voice, which is most important in personal recognition, while largely dependent upon the hereditary anatomy of the air passages, is perhaps to some extent learned from parents and teachers.

Gestures and other body movements, important for recognition at a distance or from behind, are mainly learned but certainly partly hereditary.

Body painting, hair dyeing and dressing, and clothing are of course entirely acquired, though artificial and innate colour may be difficult to distinguish from one another.

Body odour is mainly dependent upon diet and hygiene, but is almost certainly partly dependent upon the morphology and biochemistry of the skin glands.

It will be seen that we recognize people mainly by the shape and colour of the body and its parts. Most of the describable features are essentially hereditary in nature but they can be affected by the history of the individual both before and after birth. There is still much to be discovered as to the extent to which each one is determined by heredity on the one hand and by the environment on the other. Both influences are always at work upon these features and they may be difficult to disentangle; the study of their heredity belongs to the science of genetics: that of the effects of environment to ecology.

In the utilitarian business of recognizing individuals, it does not greatly matter whether what we observe is inherited or is acquired through the effects of the environment. If we are concerned about improving the environment then the acquired characteristics are primarily important, but if we want to study and distinguish individuals, the inherited ones are of major importance. This is especially the case if we want to be able to keep track of an individual over a long period of time, as for instance for police purposes. In this respect the importance of the fingerprint patterns, which are essentially hereditary, and which become fixed at an early period in antenatal life, has long been recognized.

Individuals and populations

So far we have considered solely the identification of individuals, and indeed a study of humanity must always start with individuals. However, the anthropologist is interested more in populations than in individuals. He will be concerned with environmental influences, but primarily with hereditary ones. He will wish to identify the historical and hereditary relations between populations, trying to answer such questions as whence, in Africa, the ancestors of present-day American Negroes came, or to what extent modern Jews are descended from the Palestinian Jews of Biblical times.

For such purposes characteristics are needed which are determined solely, or almost solely, by heredity and, as we shall see, it is a great advantage to make use of characteristics not merely solely determined by heredity, but also determined in a known way. Some of the characteristics so far mentioned appear to be shaped solely by heredity but for none of these directly observable features is the precise mechanism of inheritance known.

Visible and invisible characteristics

In contrast to these visible characteristics, research during the present century shows that there is a class of invisible ones, fixed by heredity in a known way at the moment of conception, immutable during the life of the individual, and observable by relatively simple scientific tests.

These are the 'blood groups', to which have been added in recent years a large number of other chemical features of the blood. The longest known and most widely used of these sets of characteristics, recognized 80 years ago, is the system of ABO blood groups, A, B, AB, and O, to one of which all human being belong, and which are familiar to all donors of blood for transfusion. And discovered, it seems, almost yesterday are the HLA or histocompatibility groups, detected by blood tests but used to ascertain compatibility in the case of proposed grafts of kidneys and other organs.

These characteristics belong to a number of genetic systems, a term defined in the next chapter, and those of any one system are inherited independently of those of all other systems. Since between fifty and one hundred such systems are now used in routine tests, the numbers of possible combinations are of astronomical dimensions and, for police purposes as an example, we are now approaching the situation where, except for identical twins, every human being will be distinguishable from every other one.

For describing and classifying populations, too, these blood characteristics have now long surpassed, and indeed largely superseded, studies of anatomical features in their usefulness to anthropologists.

2. Elementary genetics

Before we can use, or understand the use of, any inherited characteristics in the study of individuals and populations, we must know how these characteristics are inherited.

Rather surprisingly, the mechanisms of biological inheritance are essentially the same throughout the whole living world. The basic principles were discovered in the 1850s by an Augustinian monk, Gregor Mendel, growing peas in a monastery garden at Brünn or Brno in what is now Czechoslovakia. He made and published his discoveries at about the same time as Darwin published *The origin of species* but, whereas Darwin's work gave him immediate fame, that of Mendel was almost entirely ignored or forgotten until the principles were independently rediscovered about 1900, the year in which the blood groups also were discovered.

Cells and reproduction

The science of genetics is largely concerned with the inheritance of specific characteristics, such as those of shape, colour, and biochemistry. Before we look at such matters, however, we need to know the main anatomical and microscopic facts associated with the transmission of life from one generation to the next. Many forms of life consist of single separate microscopic cells, which in most cases reproduce by a process of splitting into two. Most organisms, however, both animals and plants, that are big enough to be seen by the unaided eye, consist of thousands, and in most cases millions, of microscopic cells, and reproduce by a sexual process. This involves the production by the parents of specialized reproductive cells. Two cells, one from each parent, come together and unite to form a single new cell which grows, divides into two, and continues to grow and subdivide to give rise to a new multicellular individual. The female or egg cell, known as the ovule in plants or the ovum (plural, ova) in animals, is relatively large for a cell, and contains, besides the purely reproductive part, a supply of food for the new organism.

The male reproductive cell is smaller than the female. In higher plants it is produced by the pollen grain during a somewhat complicated process which takes place within the flower. In animals the microscopic spermatozoon (pural, spermatozoa) is motile and swims to seek out and unite with the ovum. In fishes and amphibians ova and spermatozoa are in most cases shed into water and there unite. In reptiles, birds, and mammals including man, a fluid containing the spermatozoa is introduced by the male into the reproductive passages of the female.

Long before the mode of inheritance of individual characteristics was understood by anyone except Mendel, many of the microscopic details of the process of union had been worked out and were ready for incorporation into the science of genetics when the scientific world itself was ready. Every somatic (body) cell, and every reproductive cell, male or female, plant or animal, consists of a semi-fluid cytoplasm surrounding a denser nucleus. The nucleus can readily be made visible under the microscope by staining with certain dyes and, under higher powers of the instrument, made to show the details of its structure and the changes which these undergo during the reproductive process.

At a relatively low magnification the nuclei of the male and female reproductive cells are seen to unite into a single nucleus within what is, for the moment, a single cell. The new nucleus, and the cell of which it forms a part, now divides into two; each of the halves again divides, and the process is repeated almost indefinitely. In the nineteenth century all that could be seen were dead preparations representing stages in the process, but the whole process can now be watched directly and photographed by micro-cinematography.

Chromosomes

Under high magnification the nucleus is seen to contain a number of threads known as chromosomes, which are constant in number in any one species of animal or plant. The number in each of the reproductive cells, male or female, is half that in each of the body cells. The reproductive cells are said to be haploid (from a Greek word meaning single) while the body cells are said to be diploid (meaning double).

At each process of union of two reproductive cells, each such cell or gamete, one from the male and one from the female parent, contributes half the chromosomes of the new single cell, or zygote. When the zygote divides and subdivides, as described above, each resulting cell contains the full or diploid number of chromosomes. In this case too only dead 'stills' could at first be examined, but it has now become possible to see and photograph the whole active process.

In spite of general ignorance of the 'laws' governing the inheritance of body characteristics in general, it was realized before the end of the nineteenth century that the chromosomes were in some sense the carriers of such inheritance.

This development was largely due to August Weismann, a German zoologist who has never received the full recognition which the scientific world owes him for his interpretation of the facts of heredity and cytology (the science of cells). Unlike Darwin and Mendel, who made very great single discoveries but had, as we should now say, 'one-track minds', he was a man of very wide knowledge, a 'scientists' scientist' who, as Cyril Darlington put it, was able to interpret evolution to the cytologist, and chromosomes to the evolutionist.

Mendel and the mechanism of heredity

The essence of Mendel's approach to heredity was that he recorded everything in numbers, and at every stage kept the progeny of every plant, and every seed of every plant, separate. To understand his work on the common pea we must realize that the coating of a plant seed is part of, and partakes of the genetic nature of, the parent plant. But the seed itself is a complete embryo plant of the next generation, arrested temporarily at an early stage of development.

Darwin never seems to have heard of Mendel or his work. Mendel, however, though he began his work before Darwin published *The origin of species*, became in due course fully familiar with this book. Unlike Darwin, however, he was also familar with the cell theory of reproduction and development. He knew that single pollen grains fertilized single ovules, but he took the precaution of confirming this by experiment.

He found that the pea plant was fully capable of fertilizing its ovules with its own pollen. He established a number of lines of plants each of which was 'pure' for a particular characteristic such as tallness, shortness, pink or white flowers, as defined by the fact that after many generations of self-fertilization no change in this characteristic ever appeared in the progeny.

He then tried crossing pure lines. For instance, he crossed tall and short plants, using both pollen of tall plants on ovules of short plants, and vice versa. In each case all the progeny in the first generation were tall. He then grew a large number of the seeds of these plants, and in the resulting plants there was a constant ratio of almost exactly three talls to one short. He allowed the flowers of each of these plants to fertilize themselves, and then came the crucial observation. The self-fertilized seeds of the short plants produced nothing but shorts, which bred true in subsequent generations. Of the tall plants approximately one in three gave nothing but tall offspring, which continued to breed true in all subsequent generations. The other tall ones behaved exactly like their parents, giving approximately three tall to one short offspring.

The results in the first generation are explained by the statement that in peas tallness is 'dominant' to shortness (in man, however, stature is of highly complex heredity). In terms of what we should now call genes, each of the cells of the original tall plants carried two genes for tallness, symbolized by T. Those of the short plants carried two genes for shortness, symbolized by t. Thus all the body cells of the pure tall plants had the combination *TT* and all their reproductive cells, whether male or female, carried a single *T*. Similarly all the body cells of the short plants had the combination *tt*, and all their reproductive cells carried a single *t*.

Thus all the hybrid plants must have the constitution *Tt* and all will be tall, since *T* is dominant in expression to *t* (or *t* is recessive to *T*).

However, when we produce offspring from these hybrids by self-fertilization, each plant will produce equal numbers of *T* and *t* female reproductive cells, and equal numbers of male *T* and *t* reproductive cells.

A *T* female cell will have equal chances of being fertilized by a *T* or a *t* male cell, giving equal numbers of *TT* and *Tt* offspring, all of which will be tall. The *t* female reproductive cells will give rise to equal numbers of *Tt* (tall) and *tt* (short) plants. Thus the overall ratio will be 1 *TT*: 2*Tt*: 1 *tt* of which the *TT* will be true-breeding tall plants, the *Tt* tall but not true breeding, and the *tt* true-breeding short plants. We should now call the *TT* and the *tt* plants homozygous (adjective) or homozygotes (noun) while the *Tt* are heterozygous or heterozygotes. We shall see later that dominance and recessiveness are not univeral characteristics of gene expression. There are many genetic systems in man where the heterozygotes can be distinguished by suitable tests from both the homozygous types.

The new and fundamental point about these observations is that, whatever may be their expression or failure of expression in individual plants, genes for tallness and shortness remain distinct in successive generations. Every time that reproductive cells are formed the genes *segregate* separately. Many, if not most, biologists had previously thought that there was a blending not only of the expression of what we should now call genes, but of the genes themselves. We shall deal at a later stage with the much more difficult problem of characteristics such as human stature, to show how family studies which might at first sight seem to imply a blending not only of effects but of genes, are probably explicable in terms of Mendelian segregation.

So far we have used the word 'characteristics' for the separate hereditary properties of plants and animals. Now however that we have moved into the field of precise genetics we shall use the technical word 'characters' in this sense.

Mendel went on to study a further six pairs of contrasting characters in the pea, and in each case he found segregation, with the same ratios of types as he had observed for tallness and shortness.

He then embarked on a study of the effects of combining two pairs of characters in a single experiment, for instance tallness (dominant) and shortness (recessive) combined with colour of flowers (dominant) and whiteness (recessive). When a tall plant with coloured flowers was crossed with a short one with white flowers, the first generation hybrids were all tall and coloured. However, when these were allowed to self-fertilize, and the seeds grown, the next generation consisted of tall coloured, tall white, short coloured, and short white. When enough of these had been grown to eliminate chance effects, the ratios found were very close to 9 : 3 : 3 : 1, these being the only ratios which would separately yield the ratio 3 : 1 for each pair of characters taken separately. Thus not only did the genes for two contrasting characters segregate separately from one another, but they also segregated separately from those for other pairs. As Mendel pointed out, the principle could be extended indefinitely to combinations of three or more pairs of characters.

The establishment of the separate segregation of pairs of alternative characters was an essential step in the working out of the 'laws' of genetics, but there are exceptions to this rule, and Mendel was in a sense fortunate that he did not come across one of these in his basic researches, though their recognition and explanation early in the next century were necessary for the further advance of the science.

The rediscovery of genetics

The forgotten work of Mendel was rediscovered independently in 1899 and 1900 by three investigators, Tschermak, Correns, and De Vries. This rediscovery initiated a period of intensive investigation, which still continues, involving integrated studies of experimental breeding of plants and animals, microscopic examination of cells and chromosomes, and ultimately biochemistry of the most advanced nature. We shall not attempt to describe these developments, except as they are illustrated by the genetics of the blood groups and other characters used in anthropology.

The main 'work-horse' of the next generation of geneticists was the fruit fly, *Drosophila melanogaster*. It was the organism chiefly used in working out the relation between chromosome morphology and gene behaviour, initially by T. H. Morgan, and then by hundreds of other workers. It has a short generation time of ten days, it is very cheap to feed and maintain, it has hundreds of easily recognizable simply inherited characters, and it has four pairs of readily visible and distinguishable chromosomes.

This and many other small animals, as well as numerous plants, have been used in working out the fundamentals of genetics. Few doubted that the same principles would prove to apply to heredity in man or *Homo sapiens*, but the human species has, for this purpose, the disadvantages of a long generation time, and of not being available for experimental matings. However, in the long run these principles have been worked out for man, who has indeed become an important organism in the more recent elucidation of the principles of biochemical genetics. It must however be understood in what follows that in such matters as the correlation of heredity of particular characters with chromosome behaviour, the principles were already fully known before it became possible to apply them to man.

Landsteiner and the ABO blood groups

In 1900, the year which saw the rediscovery of Mendel's principles, two events took place which at first sight bore no relation to genetics. Landsteiner in Austria discovered the human ABO blood groups, and Ehrlich and Morgenroth in Germany discovered similar phenomena in goats. The impetus came not from genetics or from human biology but from the techniques of bacteriology and the phenomena of immunity to bacteria; when animals or human beings have been infected, naturally or experimentally, with certain bacteria, their blood acquires the property of agglutinating preparations of the same bacteria; when serum from such blood is added to a uniform suspension of microscopic bacteria, they agglutinate or come together in clearly visible clumps.

Blood is, of course, a suspension of microscopic red and white 'corpuscles' in a fluid. The more numerous red corpuscles are in fact cells which have lost their nuclei; they are nevertheless often referred to as 'red cells'. They consist of a membrane enclosing a solution of the red protein haemoglobin, which in the process of the circulation of the blood serves to carry oxygen from the lungs to the tissues.

The white corpuscles, only about one-thousandth as numerous, are true cells with a nucleus and cytoplasm. We shall for the moment be concerned solely with the red cells but shall return to the white cells later. The almost colourless fluid part of the circulating blood is known as 'plasma'. It contains substances which cause the blood to clot upon shedding. When the clot has separated from the shed blood the remaining fluid, differing from plasma only in the absence of the clotting principles, is known as 'serum'. Landsteiner found that, when suspensions of red cells in weak salt solution ('saline') from a number of individuals are tested separately by the addition of serum from other individuals, agglutination of the red cells takes place in some but not in other cases. Landsteiner actually distinguished three types to which a fourth was very soon added. It is not quite clear to whom should be given the credit for finding this fourth type. The four types are now named A, B, AB, and O, the symbol O indicating the absence, on the surface of the red cells, of any of the blood group substances. The symbol A indicates the presence of a substance A on cells, B the presence of a substance B, and AB the presence of both substances. We shall consider later the precise nature and origin of these 'substances'. For the present purpose they are to be regarded as 'antigens' analogous to characteristic antigens present on the surface of bacteria. It is a property of antigens that they combine with specific antibodies present in sera, the combination being demonstrated by a visible phenomenon such as agglutination. An antibody is named after the antigen with which it combines, preceded by the prefix 'anti-', the serum of a group A person contains the antibody anti-B; that of a group B person contains anti-A. Serum of a group O person contains both anti-A and anti-B, and that of a group AB person, neither. Thus the serum of any person contains as many kinds of antibodies as it can, subject to the limitation that the serum must not cause the agglutination of the subject's own red cells.

This rather complex situation is summarized in Table 1. The reader should not be discouraged by a difficulty in remembering the meaning of the new words just introduced. I well remember my own difficulty in remembering the distinction between antigens and antibodies. It is useful, as a mnemonic, to remember that anti*gen*s *gen*erate antibodies.

TABLE 1
The ABO blood groups: antigens and antibodies

Blood group	Blood group substances (antigens) in red cells	Antibodies present in plasma (or serum)
O	none	anti-A, anti-B
A	A	anti-B
B	B	anti-A
AB	A and B	none

The new discovery was presumably recognized at the outset as demonstrating a set of more or less permanent characteristics of individual human beings. The blood groups were not at first seen as having any medical consequences, nor were they recognized as hereditary. Within a few years, however, they were shown to be indispensable indicators of compatibility between donors and recipients of blood given in transfusion. Blood transfusions had been attempted on many occasions during the previous two hundred years in the treatment of haemorrhage or anaemia, sometimes with success, but often followed by rather sudden and unexplained death, now recognized as being due to incompatibility. By this one discovery transfusion was rendered almost completely safe.

Animal blood groups

We must now go back to another discovery made in the momentous year 1900, that of blood groups in goats. The investigators were Ehrlich and Morgenroth. The former is better known for his discovery of salvarsan, the precursor of all the antibiotics, and the first chemical substance able to kill an infective microorganism (in this case that of syphilis) without seriously endangering the patient. Older readers may remember a splendid documentary film, *Dr Ehrlich's magic bullets*.

The goat blood groups must have appeared even more useless than those of man, and I do not know whether veterinarians have ever applied them in transfusing goats. However, the paper describing them came to the notice of a newly qualified Polish doctor, Ludwik Hirszfeld, living in Germany. He was about to embark on a career of research, and decided to search for blood groups in dogs. Having succeeded in this, and never having heard of Landsteiners's work, he had decided to look for blood groups in man. It was at this point that he first read Landsteiner's classic paper which, far from discouraging him, only reinforced his decision to work in this field. Shortly afterwards he was thrilled to meet the great man. But he himself was destined for a career in blood group research of comparable distinction to that of his predecessor.

The heredity of the blood groups

In 1910, jointly with his Professor, Baron von Dungern, Hirszfeld showed that the blood groups were inherited as Mendelian characters. Apart from a few rare diseases these were almost the first characters shown to be inherited in this way in man. It was at this time thought that genes could exist only as sets of two (not more) alternative or allelomorphic genes (like tallness and shortness in Mendel's peas) and the results were therefore interpreted in terms of two such pairs: A and absence of A; B and absence of B; and these were thought to combine in the manner already described for tallness and shortness with colour and whiteness in peas. It was soon shown for other organisms that there could indeed be sets of three or even many more allelomorphs, but it was not until 1924 that Bernstein gave the correct explanation, which is that the blood groups are determined by a set of three allelomorphic genes, *A*, *B*, and *O* (of which any one individual carries only two). Genes *A* and *B* each express themselves dominantly with respect to the *O* gene, but *A* and *B* are not dominant with respect to one another—in the heterozygote with both

TABLE 2
Genotypes of the ABO blood groups

Genotype	Blood group
OO	O
AO, *AA*	A
BO, *BB*	B
AB	AB

A and *B* genes *both* substances A and B are present on the red cell. These relations are summarized in Table 2.

Genes and chromosomes

Very early in the twentieth century the microscopically visible chromosomes were clearly shown to be threads upon which the genes had specific places. This was most fully demonstrated in the case of the fruit fly, *Drosophila melanogaster*, which has only four pairs of chromosomes. Since each member of the half set of chromosomes present in the reproductive cell enters separately into the fertilization process we now have a clear mechanical picture of why sets of genes on different chromosomes should segregate separately, as Mendel showed them to do. But it was soon found that certain pairs of genes did not segregate independently of certain other pairs, and this was then correlated with the fact that they were situated on the same chromosome, or rather, on a chromosome of the same pair.

Taking as an example another species of pea, the sweet pea, *Lathyrus odoratus*, Punnett has shown that purple colour is dominant to red, and long pollen grains to round ones (the pollen shape is a character of the parent plant). Each pair of characters taken by itself behaves exactly as Mendelian principles would indicate. Moreover, if we hybridize purple–long with red–round, we get in the first generation nothing but the double dominant type, purple–long. However, the next generation, where simple Mendelian principles would have predicted a ratio 9 : 3 : 3 : 1 between the types, we find a ratio between purple–long : purple–round : red–long : red–round very close to 7 : 1 : 1 : 7, i.e. a great excess of the original types. There is said to be linkage between the two systems of allelomorphs, and this we now know to be due to the *loci* (plural word of which the singular is *locus*—Latin for place) being on the same pair of chromosomes.

If however the chromosomes reproduced in every generation as single entities we should expect *all* the members of this generation, not merely seven-eighths of them, to be of the two original types. Thus, in the formation of the chromosomes of the reproductive cells there has been a certain amount of interchange of genes, a process known as crossing-over, and one which has now very often been observed under the microscope.

The genetics of man tended until about 1950 to lag far behind that of other organisms. This applied particularly to the study of linkage and even more so to the microscopic study of the chromosomes, which presents particular technical difficulties, so that the examination of the chromosomes of the onion had become a regular student exercise long before the human chromosomes were even precisely counted. We shall therefore deal with these aspects of human genetics at a later stage (p. 25).

Blood groups and populations

We now return to the story of Ludwik Hirszfeld and his blood-group researches. During the First World War he and his wife, who was also a doctor, served as army doctors, and at the end of hostilities they found themselves at Salonika in north-eastern Greece. This was an important communications centre, and through it passed very large numbers of soldiers and refugees of a great many nationalities. The Hirszfelds therefore decided to carry out blood-group tests on large numbers of persons of as many races and nationalities as possible. The results are shown in Table 3. It can be seen that, while the same blood groups occurred in all the populations tested, the percentages, or frequencies, varied very widely from one to another. The Hirszfelds tested soldiers of the many European nationalities of both the contesting armies, soldiers from many parts of Asia and Africa serving in the British and French colonial forces, as well as civilians from south-eastern Europe. Apart from the Chinese and Japanese, and the indigenous peoples of Australia and America, they thus in this one pioneer paper established the main outlines of world distribution of the ABO blood groups. Moreover, unlike some of their successors, they realized the importance for reliable statistical results of testing large numbers, and as far as possible they tested 500 members of each population. Moreover, conscious of the anthropological importance of their results, they gave precise ethnic descriptions of their subjects, and of their place of origin. However, because their work did not fall into any recognized branch of science, and bore little relation to anything that had ever been published before, they had great difficulties in getting it printed. There was a very long delay in acceptance by *The Lancet*, to which it was first offered, so that it first appeared in the French journal *l'Anthropologie*.

As we have seen, Professor Hirszfeld himself was one of the joint discoverers of the Mendelian inheritance of the blood groups, but their precise mode of inheritance was not discovered until 1924, nor was it realized that the genes were in some respects more important anthropologically than the blood groups themselves. Thus the Hirszfelds' results were published in terms of the four blood groups, not of the three genes.

Genes and gene frequencies

It was largely the discovery by the Hirszfelds of the varying frequencies of the blood groups in different populations and the elucidation of their mode of inheritance that led to the establishment of the science of population genetics, and we must now look at the blood groups from this standpoint.

Mendel in 1865 had established the importance of what are now called genes as the entities which reappear unchanged in generation after generation in the process of reproduction. He himself was usually working with artificial populations of plants in which the frequencies of two allelomorphic genes were equal, that is to say that the frequency of each gene was 50 per cent, or 0.50 in terms of unity. It is however implicit in his work that even if the gene frequencies are other than 0.5, the initial frequencies will remain constant from generation to generation (apart from natural selection favouring one blood group relatively to another, and from statistical fluctuations if the absolute numbers are small—see p. 18). Let us now consider the specific example of the ABO blood groups in terms of Bernstein's three genes, *A*, *B*, and *O*, the gene frequencies, as fractions (usually expressed as decimals) of unity, being given the symbols *A*, *B*, and *O*, respectively. In Fig. 1 the sides of the square are each of value 1 (or 100 per cent), and the area of the square thus has the value of 1 square unit. The values of the gene frequencies are shown along the

	A	B	O
$\bar{A}$	$\bar{A}^2$	$\bar{A}\cdot\bar{B}$	$\bar{A}\cdot\bar{O}$
$\bar{B}$	$\bar{A}\cdot\bar{B}$	$\bar{B}^2$	$\bar{B}\cdot\bar{O}$
$\bar{O}$	$\bar{A}\cdot\bar{O}$	$\bar{B}\cdot\bar{O}$	$\bar{O}^2$

FIG. 1. Diagram illustrating gene and genotype frequencies. Each side of the square is assumed to be unity (or 100 per cent), so that the total area is also unity. The divisions of the sides of the square represent the frequencies of the genes *A*, *B*, and *O*; the areas represent the genotypes resulting from all the possible pairings of these genes.

edges of the square. We can for the purpose of argument regard the vertical distances as the frequencies of the genes in the male reproductive cells and the horizontal distances as those of the genes in the female reproductive cells. In a population in genetic equilibrium the frequencies will be equal in the two sets of reproductive cells. The frequencies of the three genes may theoretically have any values, subject to their adding up to unity, and the precise values shown in the figure must be taken as having illustrative value only.

In the following paragraphs, blood groups are represented by the roman capitals A, B, AB, O. Genes (*A*, *B*, *O*) are represented by italic capitals, as are their combinations, the genotypes (*AA*, *BB*, *OO*, *AO*, *BO*, *AB*). The frequencies (as fractions of unity) of all these three types of entity are represented by adding a horizontal bar above the symbol, e.g. $\bar{\mathrm{A}}$, $\bar{A}$, $\overline{AB}$. In contrast to the genotype, which is a theoretical concept, the observed entity, the blood group, is often called the *phenotype*.

The O genes of the fathers have only one chance of combining with the O genes of the mothers, and thus the frequency of the *OO* genotype is, subject to random statistical fluctuations, $\overline{OO}$ or $\bar{O}^2$. However the *AO* genotype may arise from the combination of a paternal *A* with a maternal *O* gene or of a maternal *A* with a paternal *O*. Thus the frequency of the *AO* heterozygote is $2\ \overline{AO}$. Similar considerations apply to the other homozygous and heterozygous genotypes.

Thus

$$\begin{aligned} \overline{AA} &= \bar{A}^2 \\ \overline{BB} &= \bar{B}^2 \\ \overline{OO} &= \bar{O}^2 \\ \overline{AO} &= 2\,\bar{A}.\bar{O} \\ \overline{BO} &= 2\,\bar{B}.\bar{O} \\ \overline{AB} &= 2\,\bar{A}.\bar{B}. \end{aligned}$$

Since the blood group A consists of the genotypes *AA* and *AO*, the frequency of blood group A is given by the equation

$$\bar{\mathrm{A}} = \bar{A}^2 + 2\,\bar{A}.\bar{O}$$

Similarly,

$$\bar{\mathrm{B}} = \bar{B}^2 + 2\,\bar{B}.\bar{O}.$$

Reverting to Fig. 1, we see that

$$\begin{aligned} \bar{\mathrm{O}} &= \bar{O}^2 \\ \overline{\mathrm{AB}} &= 2\,\bar{A}.\bar{B}. \end{aligned}$$

These results are summarized in the Hardy–Weinberg equation which states that in a given population with genes having frequencies *p*, *q*, *r*, etc. the frequencies of the genotypes are given by the expansion of the expression: $(p + q + r + \ldots)^2$

$$\text{i.e. } p^2 + q^2 + r^2 + \ldots + 2\,pq + 2qr + 2rp + \ldots.$$

These processes of calculation can be reversed and the frequencies of the blood groups or phenotypes used to calculate the frequencies of the genes. The phenotype frequencies are however subject to statistical fluctuations, and the number of phenotypes is usually greater than the number of kinds of gene. Thus the calculated gene frequencies depend to some extent on the particular phenotypes the frequencies of which are used in the calculation. To obtain the most likely gene frequencies some rather sophisticated statistical methods must be used. The processes of gene frequency calculation are described in detail by Mourant *et al.* (1976, pp. 47–61)

Since the genes are the continuing entities in any population, their frequencies, rather than those of the blood groups themselves, give the best representation of the genetic composition of the population. This can be seen, for instance, when we consider what happens when two populations mix and interbreed. In the resulting mixed population there is no simple relation between the frequencies of the blood groups in the two original populations and in the resulting mixture, but the gene frequencies are exactly proportional to those in the original populations, each multiplied by the proportion in which they entered into the mixture. Moreover, the number of genes is never more than that of the blood groups or phenotypes and may be considerably less. In the case of the ABO system it may seem to be of questionable advantage to perform elaborate calculations in order to reduce the number of variables from four observed ones to three theoretical ones, but in some blood-group systems, to be described later, such as the Rhesus or Rh system, the number of phenotypes may become quite unmanageable, and reduction to genes is essential. Thus in the rest of this book we shall, in general, express the composition of populations in terms of frequencies of genes rather than of blood groups.

One way of visualizing the meaning of gene frequencies is to imagine a group of 50 people. Each of these has on each of his or her cells two genes, the same or different, representing any given blood group system, such as the ABO system. Thus if we take one cell from each person the total number of genes will be 100 and the total number of genes of each kind will be the percentage of that gene. By dividing by 100 we get the gene frequencies expressed as fractions of unity.

When in 1918 the Hirszfelds made their classical observations on the frequencies of the blood groups in different populations no one knew exactly how these groups were inherited. We have however anticipated Bernstein's discovery by adding in Table 3 the gene frequencies calculated by modern methods.

One special advantage of using gene frequencies in the case of the ABO system to specify population composition is that the frequencies of the three genes A, B, and O is that these add up to unity, or 100 per cent. Thus any population which has been tested can be represented by the

TABLE 3

ABO blood groups and populations: the observations of the Hirszfelds and gene frequencies (later) calculated from them

Population	Number tested	Blood group percentages				Gene percentages		
		A	B	AB	O	*A*	*B*	*O*
English	500	43·4	7·2	3·0	46·4	26·8	5·2	68·0
French	500	42·6	11·2	3·0	43·2	26·3	7·4	66·3
Italians	500	38·0	11·0	3·8	47·2	23·7	7·7	68·6
Germans	ca.500	43·0	12·0	5·0	40·0	27·9	8·9	63·2
Austrians	?	40·0	10·0	8·0	42·0	27·6	9·3	63·0
Serbs	500	41·8	15·6	4·6	38·0	26·9	10·7	62·4
Greeks	500	41·6	16·2	4·0	38·2	26·4	10·7	62·8
Bulgarians	500	40·6	14·2	6·2	39·0	27·0	10·8	62·2
Arabs	500	32·4	19·0	5·0	43·6	20·9	12·8	66·3
Turks	500	38·0	18·6	6·6	36·8	25·6	13·5	60·9
Russians	1000	31·2	21·8	6·3	40·7	20·9	15·2	63·8
Jews	500	33·0	23·2	5·0	38·8	21·4	15·4	63·3
Malagasies	400	26·2	23·7	4·5	45·5	16·8	15·3	67·9
Negroes (Senegal)	500	22·6	29·2	5·0	43·2	14·9	18·9	66·1
Annamese	500	22·4	28·4	7·2	42·0	16·0	19·7	64·3
Indians	1000	19·0	41·2	8·5	31·3	14·9	29·1	56·0

frequencies of only two genes, such as *A* and *B*, and the population can thus be represented by a single point on an ordinary two-dimensional graph. Ideally, in the interests of mathematical symmetry, graph paper ruled in equilateral triangles should be used, but most people can follow rectangular graphs more easily.

Causes of change in blood-group frequencies

We have seen that blood group frequencies tend to remain constant from one generation to the next in any one reproducing population. However, the great diversities of frequency which we find even among closely related populations indicate that changes in frequency, though possibly very slow ones, must constantly be taking place.

Mutation

The initial source of such diversity is mutation, the substitution, at some stage in the production of a reproductive cell, of one gene by one of its allelomorphs. Such a change may perhaps take place spontaneously, but it may be provoked by chemicals or by radiation. It is, however, a rare event, any one gene being affected in only about one in a million reproductive cells. Thus mutation cannot by itself account for any appreciable change in the gene frequencies in a population.

Natural selection

A mutated gene, except by very rare chance, will appear in any substantial number of members of a population only if the substitution gives its carriers some advantage over those of the unchanged gene, such that they tend, in the current environment, to produce more offspring than their fellows. This is one example of natural selection, but a commoner example is found when a population suffers a change of environment and the balance of reproductive advantage between genes already present is thereby altered.

Genetic drift

Natural selection may occur in a population of any size, but another cause of change in gene frequencies, genetic drift, operates perceptibly only in small populations of less than about a thousand. In any population individual families may show chance variations in the frequency of offspring of different types. For instance, a pair of *AO* and *OO* parents should have equal numbers of *AO* and *OO* children, but they may by chance have other proportions, or even children all of one of the two types. In a large population these variations will balance one another, and the overall numbers will be very nearly equal, but in small populations random events in individual families will substantially affect overall ratios. Such effects may be reversed or enhanced in subsequent generations, but the long-term effect, so long as the population remains small, will be a gradual erratic drift of gene frequencies. However, if circumstances improve, so that the total population is able to increase considerably, the chance results of genetic drift will be perpetuated, since drift will almost completely cease to occur in the larger population.

Another difference between natural selection and genetic drift is that the former tends to affect the genes of only a few systems, whereas with the latter all systems have equal chances of being affected.

The founder effect

A phenomenon closely related to genetic drift is the founder effect. This operates when a small number of persons, drawn from a larger population, migrates to a new habitat and there breeds in isolation and increases in numbers. Because of their small numbers it is unlikely that the migrants will be a representative sample of the population from which they are drawn and thus, when they multiply, the migrant population will show gene frequencies different from the ancestral one. Founder effects are essentially the same as the results of genetic drift, and the statistics of the two processes are basically the same.

Linkage equilibrium

In systems of several closely linked genes, such as the Rh system, it is now thought that the establishment of equilibrium between the genes at the several loci is of anthropological importance. The process will be described after consideration of the genetic systems concerned (pp. 129–30).

Blood-group genes and populations

In Fig. 2 gene frequencies calculated from the observations of the Hirszfelds have been plotted. It will be seen that there is a clear separation between different 'races' and quite a marked one between the peoples even of adjacent countries. A plot for these same populations, based on more recent and much more numerous observations, would not differ greatly from the results of this pioneer effort. In particular it would still show similar rather high frequencies of gene *A* in Europe, and the progressive increase of *B* from west to east in Europe and Asia. The finer details of ABO distribution, already clearly anticipated in the Hirszfelds' work, will be much more fully considered in subsequent chapters. Among important populations not studied by the Hirszfelds, North American Indians have only the *A* and *O* genes, the Indians of

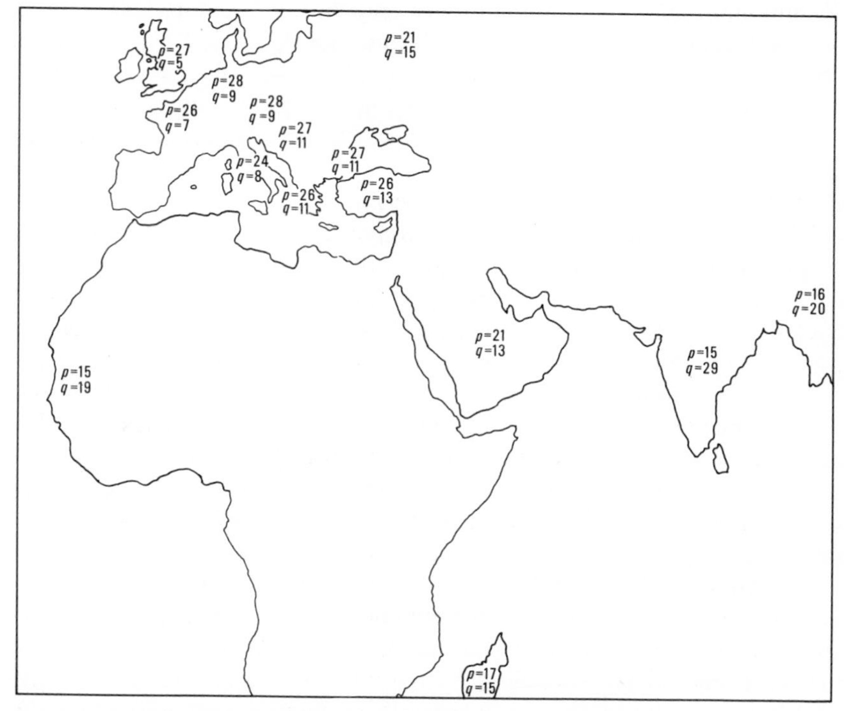

Fig. 2. Map of parts of Europe, Asia, and Africa showing percentage frequencies of genes *A* (p) and *B* (q) in the indigenous populations, calculated from the observations of Professor and Mrs Hirszfeld (Table 3). To avoid confusion frequencies of gene *O* (r) have been omitted. They are such that for each population $p+q+r = 100$.

Central and South America are nearly all of group O. Among the Australian Aborigenes, too, the *B* gene is absent except in the extreme north-east.

The pioneer work of the Hirszfelds initiated the world-wide collection of data on the frequencies of the ABO blood groups. Up to 1970 results had been published on about 15 million individuals from populations all over the world. The results of these surveys will be analysed in subsequent chapters.

The MN and P blood groups

Up to 1927 the term 'blood groups' meant simply the ABO groups, for few people had any idea that there could be others; one of these few was Landsteiner; he and Levine injected rabbits with red cells—each rabbit with cells from one person. The sera of these rabbits, after suitable treatment, were found to contain any one of three antibodies, each of which agglutinated some but not all human red cells; the substances assumed to be present on the red cells were given the symbols M, N, and P, giving rise to agglutination of the cells by the respective antibodies anti-M, anti-N, and anti-P (The red-cell substance and the antibody are now known, for rather complicated reasons, as P_1 and anti-P_1.) The P_1 substance on the cells was found to be inherited on simple Mendelian Lines, P_1 behaving as a dominant to the recessive now known as P_2, so that persons having one or two P_1 genes had red cells which were agglutinated, whereas those of persons with two recessive P_2 genes were not. The inheritance of P_1 and P_2 is independent of that of the ABO groups. Many of the available anti-P_1 reagents are somewhat unreliable, so that it is difficult to distinguish minor variations in frequency with any certainty. Broadly speaking, however, the P_1 and P_2 genes have frequencies of about 50 per cent each in European populations. The P_1 gene reaches a frequency near 90 per cent cent in African peoples, but frequencies are low in the Far East.

The case of M and N is more complicated—the substances M and N on the red cells are inherited by means of two allelomorphic genes *M* and *N* at a single locus (independent of ABO and of P_1P_2) but there is no dominance or recessiveness, so that while the cells of *MM* homozygotes are agglutinated by anti-M and those of *NN* homozygotes by anti-N, those of *MN* heterozygotes are agglutinated by both antisera.

Because anti-M, anti-N, and anti-P_1 occur only extremely rarely in human serum there is almost no danger of their causing trouble in blood transfusion. Thus they have been recognized as of little importance in medicine, and testing has largely been left to geneticists. Thus data on the distribution of the M, N, and P_1 groups has built up only very gradually over the years. On the other hand the importance of the ABO groups in transfusion has had the indirect effect of providing, through medical laboratories, a vast body of data on their anthropological distribution. A large number of other genetically independent sets of blood groups have since been discovered, but if we are to understand the genetics of these we must enter more fully into some details of general genetics, including cytogenetics.

Chromosomes in reproduction

We have seen that in reproduction in higher organisms the ovum and the spermatozoon come together to form a new cell. Each contributes an equal number of chromosomes and, with one exception to be dealt with later, the chromosomes so contributed are paired, one coming from the spermatozoon and one from the ovum. The members of a pair are of equal length and of closely similar structure.

During the first half of the twentieth century it was shown, by a combination of breeding experiments with optical microscopy of the chromosomes of a great variety of organisms, that each chromosome behaves as a string of genes in a fixed linear order, and that the paternally and maternally contributed genes for a single character, such as flower or hair colour, occupy corresponding places on a pair of chromosomes which are of equal length and show special pairing behaviour at certain stages of the reproductive process.

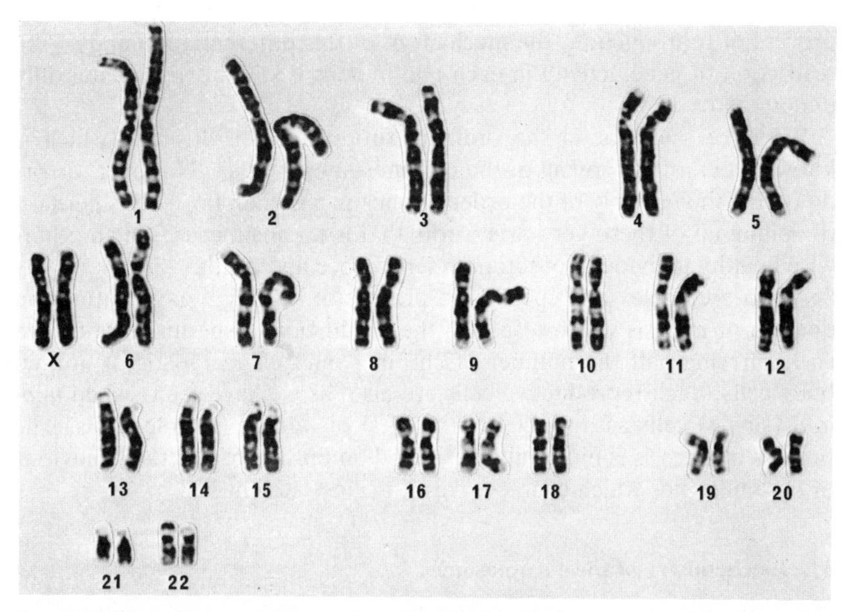

FIG. 3. A karyotype of a normal female, stained to reveal the banded structure of the chromosomes.

We shall first look at the processes of cell reproduction and sexual reproduction of the whole multicellular organism as seen under the microscope. We shall then look at the extraordinarily full explanation of these processes in molecular terms which has been built up during the third quarter of the present century.

Figure 3 shows the 23 pairs of chromosomes present in the initial human cell or zygote resulting from the conjugation of ovum and spermatozoon. In the subsequent reproduction of this initial cell, all the 46 chromosomes are reproduced exactly at each cell division, with only the rarest of exceptions. At first not only the nuclei with their chromosomes, but the whole cells, are reproduced exactly. Later the cells become differentiated in function into skin cells, nerve cells, liver cells, and a multiplicity of others, but the nucleus of every cell contains the same set of chromosomes, and the information needed for producing any cell or combination of cells. Indeed, in plants, though not at present in animals or man, it is possible under suitably controlled conditions to reproduce the whole organism from a single cell.

It is clear that each set of chromosomes contains complete instructions for the building up of all the organs of the body in their correct relation to one another; these instructions include those applying to properties (such as colour) of certain tissues, in which individuals differ from one another. In animals, however, and to a lesser extent in plants, the cells also acquire certain properties which cause skin cells to divide into skin cells, liver cells into liver cells, and so on. In spite of our understanding of the main process of reproduction, the mechanism of this differentiation and of the restriction of gene activity in each line of cells is still far from being fully understood.

We have said that in the ordinary process of cell division (which is known as *mitosis*) copying of the chromosomes is exact. However, errors do occur, though only of the order of once in a million times; it is perhaps the piling up of these very rare errors that is the main cause, in an otherwise healthy individual, of ultimate senescence and death.

A more complex and specialized process of cell division is maturation division or meiosis which results in the production of spermatozoa or ova each carrying half the number of chromosomes characteristic of normal body cells. Such reproductive cells are also, as we have seen, called *haploid* (single) cells as distinct from normal or *diploid* (double) cells. The process of meiosis is more fully described on pp. 25–6, and the behaviour of chromosomes which characterizes it is illustrated in Fig. 3.

The biochemistry of the chromosomes

The cell nucleus was at an early date the subject of chemical analysis, which became more and more sophisticated with the progress of biochemistry, and with the realization of the function of the nuclei, and of their contained chromosomes and genes, in cell reproduction.

At the same time it was becoming clear that the primary products of gene action were proteins, and in many cases enzymes (the proteins which catalyse most of the chemical reactions in the body).

The blood groups themselves, however, are not, as was once thought, primary gene products. Their complex molecules are built up in several stages through the action of enzymes which are directly produced by the genes.

The nucleus, like the rest of the cell (the cytoplasm) and non-cellular tissues, contains protein, mainly of a type known as histone, the function of which is not quite clear, but the nucleus is particularly notable in its content of phosphate, of carbohydrates compounded of pentose (5-carbon) sugars, and of certain organic bases containing a high proportion of nitrogen and known as purines and pyrimidines. Each chromosome is in fact an enormously long chemical molecule, consisting of alternating sub-molecules of phosphate and of the 5-carbon sugar desoxyribose. All the phosphates and all the sugars are identical, so that there is no basis in them for differences between chromosomes or between their constituent genes. However, attached to each phosphate is a side-chain of purine or pyrimidine. Two kinds of purine and two of pyrimidine are involved. The purines, guanine and adenine, each with 5 carbon and 5 nitrogen atoms, are larger than the pyrimidines, thymine and cytosine, with 5 carbons and 2 nitrogens, and with 4 carbons and 3 nitrogens respectively.

It had long been known that the chromosomes , and presumably their constituent genes, had a spiral structure, and it was strongly suspected that the numbers and arrangement of the purine and pyrimidine bases had much to do both with the specific activity of the genes and with their reproduction.

The decisive discovery was made in 1953 by Watson and Crick, who showed that the chromosome consisted of a double helix, that is to say a pair of spiral molecular chains, and that because of the shape, size, and chemical activity of the bases, adenine in one helix was always opposite to thymine in the other, and guanine was opposite to cytosine. It was strongly suspected at this stage, and was soon proved conclusively, that the nature and sequence of the purine and pyrimidine bases in a chain was a code for the nature and sequence of the amino-acids in the proteins which are the final product of genetic action, and are the building material of cells and tissues.

The nature and sequence of the purines and pyrimidines in one of the spiral chains thus determines the nature and sequence of those in the opposite spiral. Thus if the first and second chains are stripped apart a duplicate first chain can be built up upon the second chain, and upon this duplicate first chain a new second chain can be built. These processes can theoretically go on an infinite number of times, and thus lead to the continuing precise reproduction of the structure of the initial chromosome and its gene. The amino-acids in the protein which is to be synthesized are each determined by a sequence of three bases, and this coding of the amino-acids is now almost completely deciphered.

It is most surprising to realize that the whole of life and reproduction as we know them depend upon something analogous to a literature of letters, words, and sentences evolved by simple microorganisms at least one thousand million years ago. There is an alphabet of four signs, two purines and two pyrimidines, which form words of three letters symbolizing the amino-acids, and these are built into sentences: the proteins—and so we might go on, for it is estimated that the information built into the chromosomes of every cell of a higher animal or plant is comparable to the contents of a large library.

There is in fact a redundancy in the coding, for there are 64 possible sequences of four bases taken three at a time, while only 20 amino-acids ever appear in proteins. It is now known that certain amino-acids can be coded in several ways, and that a few codes are used for other purposes; and perhaps some are meaningless.

We have seen, in summary form, how genetic information is stored and replicated in the chromosomes of the nucleus, and that this is basically very simple. The translation of this information into protein synthesis is much more complicated and involves another type of sugar–phosphate–organic–base chain. The sugar in this case is not deoxyribose, but plain ribose, and the pyrimidine base which pairs with adenine is not thymine but uracil.

A length of single separated DNA chain, corresponding to a single protein, delimited by certain codings, is separated from its fellow of the double helix, and is now the site of synthesis of a new double helix, of which the second spiral chain is to consist of ribonucleic acid, and more specifically, of messenger ribonucleic acid, or messenger RNA, synthesized of the constituents just mentioned. It is then stripped off and, apparently because of its RNA basis, is able to leave the nucleus for the cytoplasm where it takes part in the synthesis of numerous (identical) molecules of protein, one at a time. The process is not fully understood, but it takes place in a body known as a ribosome which runs along the RNA chain, rather like the 'zipper' of a zip fastener, taking up separate molecules of amino-acid from the cytoplasm and incorporating them in a polypeptide chain. Thus there is, in front of the ribosome, the untranslated part of the RNA chain, while emerging from behind it is the translated part of the chain, and the translation in the form of a growing polypeptide chain which will, when complete, be a protein molecule. Each kind of amino-acid is added to the chain by a special type of RNA known as transfer

RNA, which recognizes the code for that amino-acid and causes it to be added to the chain.

Back to chromosomes

We must now, as it were, alter the setting of our zoom-lensed microscope from its adjustment to molecular chains and events, and again look at the chromosomes as single entities. The molecular processes which we have described are those of mitosis, the duplication of the diploid cells of the animal or plant body. We are now to look at meiosis, the production of reproductive or haploid cells by means of 'reduction division' in which the number of chromosomes is halved.

The parent cells of the reproductive cells, whether male or female, are diploid, that is, they have a full or double set of chromosomes like all the other body cells, and are the products of a long succession of cell divisions by mitosis. We must now study the complex series of events that leads to halving of the chromosome number: in man this is from 46 to 23. Let us look at what happens to one of the 23 pairs of chromosomes. One member of the pair is identical with that orginally received from the mother and one with that from the father.

The next stage is peculiar to meiosis. Each of the 23 chromosomes of paternal origin sets itself closely alongside the corresponding or homologous chromosome of maternal origin (Fig. 4). Each chromosome then divides into two objects known as chromatids, which are essentially new chromosomes, giving rise to a closely set four-stranded structure (Fig. 5). At various levels in the quadruple strand two of the four strands cross over one another and exchange segments. In Fig. 6 only a single crossover is represented, but in general numerous crossovers take place in each set of

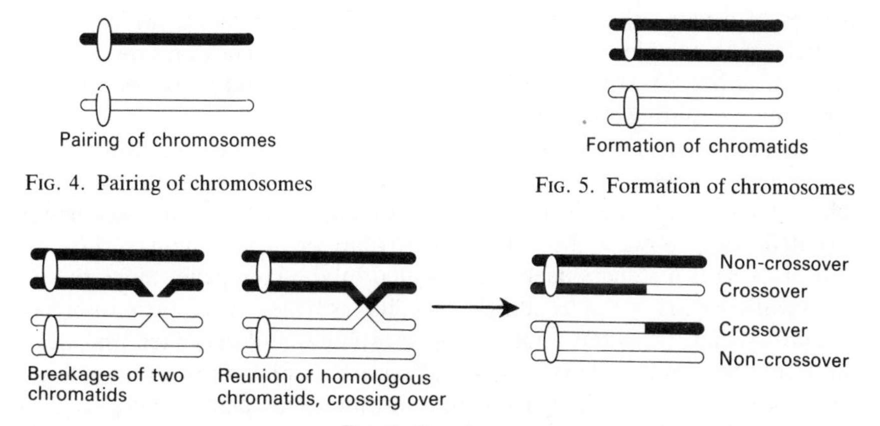

FIG. 4. Pairing of chromosomes

FIG. 5. Formation of chromosomes

FIG. 6. Crossing over

four strands. At any given level in the set only two out of the four cross over one another, but not necessarily the same two at each level. Thus in a long chromosome there will ultimately be four strands each with some segments of paternal and some of maternal origin.

The cell, which at this stage has 92 chromosomes, or 23 of the four-stranded structures just described, now divides, each resulting cell carrying the normal number of 46, but each chromosome now has a mixture of paternal and maternal genes, and these will be unlike in each of the still closely apposed 23 pairs. Each of the two cells now divides again, giving rise to two cells (or four in all) each with only 23 chromosomes. In the male each of these cells ultimately becomes a spermatozoon but in the female only one survives as an ovum.

Linkage

One of the genetic consequences of the morphological events just described is that of linkage.

Genes which are on separate chromosomes in the haploid or reproductive cell segregate separately so that, for instance, a maternal gene on one chromosome may be accompanied by either a maternal or a paternal gene on another. Owing to crossing-over, the same is true of genes remote from one another on a single chromosome. However, genes which are close together on a single chromosome may remain associated through the whole process of meiosis, and the closer they are, the more likely are they to remain together, or to be 'linked'.

The sex chromosomes

There is one exception to the close similarity between the two chromosomes of a pair—this concerns the X and Y chromosomes which determine sex. A female has two X chromosomes and passes on one of these to each of her offspring. A male has an X and a Y and passes on an X to his female offspring and a Y to the males.

The X chromosome is much longer than the Y and carries genes for a very large number of recognizable characters. Though the short Y chromosome undoubtedly carries a great many genes, very few characters are definitely known to be determined by any of these. The most important of these is maleness itself, determined by the H-Y gene. Since the inactive allele of H-Y is exceedingly rare, it was believed for more than half a century that it was the whole Y chromosome that determined maleness, and only the study of very rare sex abnormalities has shown that a single gene is responsible. Though the sex chromosomes pair at meiosis, it is still uncertain whether any crossing-over takes place between the X and the Y in the male. At meiosis in the female, however, crossing-over does take place between the two X chromosomes. Sex itself therefore behaves in normal cases as though determined by the whole X and Y chromosomes.

The characters determined by genes on the X chromosome are known as sex-linked or, more correctly, as X-linked. Because of the special connection of this chromosome with sex determination, they behave as linked to sex, but since crossing-over takes place at meiosis in the female, only those genes which are very close together on the X chromosome remain linked to one another through several generations.

Most of the known X-linked systems express themselves as pairs of contrasted characters consisting of a full complement on the one hand and a deficiency or absence on the other, of some important enzyme such as a blood- clotting factor. In the female they thus become manifest only when the deficiency is transmitted by both parents, and so behaves as a recessive, but in the male, with only one X-chromosome, all the genes upon it are fully expressed.

Very few of the blood-group and other marker genes used in population studies are X-linked, and those that are will be considered later. However, an understanding of linkage in general is necessary for the consideration of some of the other blood groups, determined by genes on the autosomes, which is the general name given to the chromosomes other than the X and Y.

The rest of the blood groups

The ABO, MN, and P blood-group systems were discovered, and indeed applied to population studies, long before the remaining systems, which were mostly found fairly rapidly after one another, beginning in 1940. To understand them, and especially the very important Rhesus system, we need a considerable knowledge of general genetics, and especially of the theory of linkage, which is why a general genetic section has been inserted between the accounts of the first three systems, and those of the rest which are now to be described.

The Rhesus blood groups

The Rhesus or Rh blood groups were discovered in 1940 by Karl Landsteiner, this time in collaboration with Alexander Wiener. The name was given because the serum originally used in the research was made by immunizing rabbits with rhesus monkey cells, but today human serum is used exclusively.

At first only one antigen was found, now usually known as the D antigen, which behaves as a dominant character. Persons with two *D* genes, one from each parent, or with one *D* and one of the allele *d*, carry the D antigen on their red cells, whereas homozygous *dd* persons do not.

The Rh system was soon shown, by Levine and his colleagues, to be of greater medical importance than the ABO system, for it showed the cause, the cure, and ultimately the way to prevention, of the hitherto puzzling disease now known as haemolytic disease of the newborn. When an Rh-negative or *dd* woman is married to an Rh-positive (*Dd* or *DD*) man, their children are likely (and with a *DD* father they are certain) to inherit the *D* gene and the D antigen. Under certain conditions red cells may escape from the circulation of the unborn child into that of the mother, and she is likely to respond by making an antibody, anti-D, which, being in solution, readily goes back across the placenta into the circulation of the infant. In a first pregnancy this usually happens too late to cause trouble, but if the mother, now with anti-D in her plasma, has a second D-posivite baby, the anti-D attacks the fetal D-positive red cells and destroys or haemolyses them (hence the name of the disease) causing anaemia which may be very severe, as well as jaundice due to breakdown products of the red cells. The treatment of this formerly highly fatal disease is transfusion with blood from an Rh-negative (i.e. D-negative) donor.

Medically speaking this is nearly the whole story, but genetically and anthropologically it is not. It soon became clear that the system is a highly complex one genetically. There has never been serious disagreement about the serological observations, but the genetic explanation of them was for many years a subject of heated controversy. Even today no one theory explains all the facts, but nearly all of them can be explained, as was first done by Professor R. A. Fisher and Dr R. R. Race, in terms of three loci, very close together on a single chromosome which is not that determining the ABO or that determining the MN blood groups. Each locus may be occupied by one of a pair of alleles, one by *C* or *c*, one by *D* or *d*, and the third by *E* or *e*.

When the symbols were allotted it was known (or strongly suspected) that three loci were involved which were very close together, but their se-

TABLE 4

Rh gene complexes (haplotypes) showing the CDE (Fisher) notation and the usual abbreviated British notation based on that of Wiener

Fisher's notation	British notation
CDE	R_z
CDe	R_1
cDE	R_2
cDe	R_o
CdE	r^y
Cde	r'
cdE	r''
cde	r

rial order was not known, so that the letters were allotted, in effect, in random order. It now appears probable that the real order is *DCE* but in this book we shall retain the familiar alphabetical order.

Since there are three loci and two possible genes at each, eight chromosomic combinations are possible, as shown in Table 4. Various general terms have been used for these three-gene combinations but, for the sake of uniformity with other closely linked systems, the term haplotype will be used here. All the eight have been observed, but *CdE* is extremely rare in all populations. The haplotypes *CDE, CdE*, and *cdE* are somewhat rare in all populations, but *CDe, cDE, cDe*, and *cde* are, in general, fairly common. The proportions of them, however, vary widely and thus these four haplotypes provide a wealth of anthropological information.

The expansion of the MN system

When the MN system was first discovered it appeared rather uninteresting. It was of virtually no medical importance, and the frequencies of the *M* and *N* genes were closely similar, each near 50 per cent, in most populations available for testing. Only the American Indians, with much more *M* than *N*, relieved this uniformity. The discovery of the S and s antigens in 1947 and 1951 came at a fortunate time, for a model of closely linked genes in the Rh system had been very thoroughly studied, and so it was soon realized that here again closely linked loci were involved, two in number, one determining the alleles M and N and the other S and s. Thus in the MN system we now have the possibility of four haplotypes *MS, Ms, NS*, and *Ns* and, as we shall see, this expansion of the system greatly enhances its anthropological value.

A considerable number of other antigens are now known to be determined by genes closely linked to *MN* and *Ss* but only one of these, the Henshaw or He antigen, has any great anthropological value, for it appears to be totally limited to populations of African ancestry. The hypothetical allele of the *He* gene, which we may call *he*, has not yet been shown to give rise to an antigen.

The Kell system

The Kell system of blood groups may be looked upon as a pale shadow of the Rh system. It similarly is determined by three closely linked loci, each of which can be occupied by one of a pair of allelic genes, *K* and *k*, Kp^a and Kp^b, and Js^a and Js^b. The *K* gene, like *D* in the Rh system, gives rise to an antigen to which mothers not infrequently become immunized, so that their infants suffer from haemolytic disease of the newborn. The *K* gene and its antigenic product are almost entirely confined to the white or caucasoid peoples, and Js^a to the negroid or African ones.

The Duffy system

The Duffy system of blood groups is genetically simple, being controlled by four allelic genes at one locus, of which only three are sufficiently common to be of anthropological and medical significance. Since the letter D is used for the principal gene of the Rh system the last two letters of the name Duffy are used in the gene symbols, Fy^a, Fy^b, and Fy^4. The Fy^4 gene is very rare outside Africa and very common within that continent. Not only is it therefore an important anthropological marker, but, as described on pp. 121–2, it provides one of the few cases where we think we know, in terms of natural selection, why one population differs from another in their blood-group frequencies.

The Diego system

While Africans, or Negroids, have a considerable number of blood group marker genes peculiar to themselves, east Asians, or Mongoloids, have only one known one, the Diego or Di^a gene. The allelic Di^b gene is, even in Mongoloid populations, the commoner of the two. The Di^a gene is interesting as being possessed in common by east Asians and American Indians, and this is one of the indicators of the Asian origin of the latter.

The Gerbich system

The Gerbich blood group system depends upon two allelomorphs, Ge^a and the amorph *Ge*. Nearly all persons outside New Guinea are homozygous for Ge^a, but in New Guinea, and especially in the ancient inland populations, the *Ge* gene has a substantial frequency of up to as much as 80 *per cent*. Locally therefore it is an important marker gene.

The ABH secretor system

Genetically independent of the ABO blood groups is the secretor system which comprises two allelomorphic genes, *Se* which causes secretion in saliva and other fluids of the antigen or antigens corresponding to the individual's ABO group, and *se* which, in the homozygous condition determines non-secretion. Heterozygotes are secretors. There are wide variations in the frequencies of the two genes, which could therefore be of considerable anthropological value. Also the secretor and non-secretor states have significant associations with certain diseases. Natural selection is therefore thought to be of considerable importance in determining the frequencies of the genes.

The Gc system

The Gc groups of the plasma are the expression of genetically controlled variants of a protein belonging to the α-globulins, which is the carrier of vitamin D. There are numerous alleles of which only two, Gc^1 and Gc^2, are relatively common. They are distinguished by methods based on electrophoresis. The distribution of the alleles and the possible relation of this to sunshine are described on p. 123.

The protease inhibitor system

The body contains a number of digestive enzymes known as proteases which break down proteins into their component amino-acids. These are on the whole confined to the digestive tract and other places where they are needed, but they do to some extent get through into the blood plasma, where they tend to break down the tissues of the body indiscriminately. This tendency is counteracted by the presence in the plasma of a protein which functions as a protease inhibitor. This exists in a large number of forms controlled by a set of alleles, and having different degrees of inhibitory activity. There is one with extremely low activity, and there also probably exists an amorph gene which produces no activity at all.

The glucose 6-phosphate dehydrogenase system

Inside the boundary membrane of the red cell is a solution containing a great variety of proteins, including haemoglobin and numerous enzymes many of which have several allelic variants. The variants often show quantitative differences in efficiency. One enzyme with such variants is glucose 6-phosphate dehydrogenase. This protein is needed for maintaining the integrity of the red cell. The gene concerned is carried on the X chromosome, so that males have only one such gene but females have two. In males with a deficient gene or in females with two such genes there is a tendency for the red cell to break down if the individual consumes certain drugs, or the common broad bean (*Vicia faba*).

The phosphoglycerate kinase system

The red cell enzyme phosphoglycerate kinase, like glucose 6-phosphate dehydrogenase, is involved in maintaining red-cell viability, and also like it, is controlled by genes at a locus on the X chromosome. Most populations everywhere have only the PGK^1 allele and it is almost solely in New Guinea and the adjacent islands that other alleles are found, and the system becomes of anthropological importance.

The red-cell acid phosphatase system

The phosphatases are enzymes which catalyse the transfer of the phosphate radical (PO_4) between the inorganic ionic state, as in sodium phosphate, and attachment to an organic radical, as in glucose 6-phosphate. There are numerous kinds of phosphatase in the body. One of these, red-cell acid phosphatase, is present in red cells and is most active under acid conditions. It has several genetic variants, distinguished by their electric charges, and hence by their speeds of migration during electrophoresis. They also differ somewhat in the strength of their enzymic activity. The variants P^a, P^b, and P^c are present in most populations with greatly varying frequencies, while P^r is found almost solely in southern Africans, and especially in the Khoisan peoples.

The malate dehydrogenase system

The red cell enzyme malate dehydrogenase catalyses the oxidation of malate to oxaloacetate. It is not sex linked. Several alleles are known but most populations are uniformly homozygous for one allele $SMDH^1$. Most other alleles are very rare but one is present in high frequency in numerous New Guinea populations.

The haemoglobins

Haemoglobin is the protein which carries oxygen from the lungs to the tissues. Its complex molecule includes four polypeptide chains (i.e. chains of amino-acids produced by genes, and characteristic of proteins in general). In any one molecule there are two identical alpha chains and two identical beta chains. It is variants of the beta chains which are important physiologically and anthropologically. Because the functional activity of haemoglobin is so important for efficient living, selection has almost completely weeded out the genes for all but one kind of beta chain, that which characterizes haemoglobin A. With a very few exceptions all other types of beta chain are exceedingly rare. However, in Africa and elsewhere, for reasons discussed on pp. 125–6, the gene for haemoglobin S (sickle-cell haemoglobin) is fairly common. This haemoglobin behaves almost normally in the oxygen-saturated state, but when it gives up its oxygen it becomes insoluble in the red-cell fluid, forming crystals of a sort, which distort the cell and tend to cause its disruption. As explained on pp. 125–6, it is subject to a very important process of natural selection.

Other beta variants are that for Haemoglobin E, fairly common throughout south-east Asia, and Haemoglobin D, present in the Gujarat region of India and the Punjab. Possible selection processes favouring these are discussed on p. 126.

The thalassaemias

Closely related to the haemoglobin variants are the thalassaemias due to genes which suppress (or fail to activate) the synthesis of the alpha or the beta chain of haemoglobin. Beta thalassaemia is the only one of interest in population studies. Heterozygotes with the normal allelic gene suffer from only a mild degree of anaemia but appear to have a raised resistance to malaria, while homozygotes suffer from severe anaemia and if untreated mostly die in childhood. The formal genetic situation is thus closely parallel to that involving haemoglobin S, and a similar process of natural selection seems to take place in malarial environments.

The histocompatibility system

Besides the red cells, the blood contains a number of different kinds of white cells, or leucocytes, important in various ways in controlling infective diseases. The lymphocytes are one of these kinds. Their main function is the production of antibodies, but they contain a variety of genetically controlled antigens, detectable by their reactions with specific antibodies naturally present in certain human sera. These antigens were at first regarded as pathological curiosities, until it was found that they were present in other tissues and that compatibility with respect to them was a major requirement for the survival of kidney grafts and other tissue and organ grafts. This discovery stimulated world-wide researches into their genetics, and it is now known that they are the products of a set of alleles at five loci, closely linked on a single chromosome (No. 6). Because of this close linkage the genes, and the corresponding antigens, are passed from one generation to the next in groups of five. They are however not so closely linked as are the linked genes of the blood group systems MNSs, Rh, and Kell and, unlike the situation with these blood-group systems, the rates of crossing-over between them are known, being of the order of 1 per cent per generation. Crossing-over should in the long run produce a state of linkage equilibrium, such that a particular allele at one locus should be linked to a pair of alleles at another locus in the same ratio as the total frequencies of these alleles. This is however often found not to be the case, and many workers think therefore that certain combinations or haplotypes are selectively favoured. This and other possible causes of such disequilibrium are discussed elsewhere (pp. 128, 129–30).

A great deal of work has been done on the frequencies of the HLA alleles and haplotypes in populations and it is clear that there are wide variations in frequency in different populations, so that the system will, when more work has been done, become of great importance anthropologically. Up to the present most of the surveys reported refer to caucasoid populations, because tissue and organ grafting is mainly confined to Caucasoids.

However the main biological importance of the histocompatibility antigens is proving to be not their relation to grafting but the highly significant associations which exist between particular alleles and haplotypes and a large number of diseases, mainly diseases with an immunological basis. Many of these associations are so close as to have an important bearing on the causes of the diseases, which in many cases were formerly obscure. These associations are discussed on pp. 127–8.

The phenylthiocarbamide taster system

To some people the chemical substance phenylthiocarbamide tastes intensely bitter while to others it is almost tasteless. The ability to taste this and a number of other chemically related substances is inherited. There are two genes, *T* for tasting and *t* for non-tasting. *T* is dominant in expression over *t*. This system was one of the first genetically simple systems to be discovered in man; it has therefore been widely applied in population studies, and there are considerable differences between populations in their taster frequencies.

Phenylthiocarbamide, and all the other synthetic substances which elicit the taster-non-taster phenomenon, have thyroid inhibitor properties, and some of them have been used therapeutically for this purpose. Certain substances of this class also occur in vegetable foods such as cabbage, and epidemics of thyroid disease have many times been caused by excessive consumption of these foods, such as in times of famine. It is not surprising that associations with tasting and non-tasting have been found in thyroid diseases.

Iodine forms part of the molecules of the thyroid hormones, and iodine deficiency is well known as a cause of thyroid disease. Attempts have therefore been made to explain frequency differences in terms of natural selection related to the amounts of iodine and of thiocarbamides in the normal diet (pp. 128–9).

Visible body characters

Directly observable characters, such as the shape, size, and colour of the body and its parts, were formerly the only means of classifying individuals and populations. They have several disadvantages for this purpose, particularly their complex inheritance, and the fact that almost every character is influenced both by heredity and environment.

Though they have for these reasons largely been superseded for purposes of classification by the blood groups and other hereditary blood characters, it must be borne in mind that they are still the means by which in everyday life we recognize people, and that quantitative observations of them are the only means we have of comparing skeletal material, and observations made on the living before the discovery and application of the blood groups, with people living today. They also present very clear indications of probable natural selection in relation to the environment. They must therefore continue to be observed with as much precision as possible.

It would be a great advantage if their heredity could be more fully understood. It is now clear that almost any one single character, such as stature, is the effect of genes at a considerable number of different loci, so that they are known as polygenic characters.

Attempts have been made, mostly long ago and with little success, to unravel the separate effects of the individual genes involved. The recent very rapid advances made in human genetics as a whole may however be opening the way for a new approach to the genetics of these characters. If their genetics could be understood this would make a major contribution to human biology and classification.

3. Africa

We have now examined something of the nature of the blood groups and some other blood factors and seen how they are inherited. In particular we have seen that basically the proportions of any particular set of factors, and those of the genes on which they depend, tend to remain constant from generation to generation, and any changes due to such processes as mutation, natural selection, and genetic drift are likely to be very slow, so that two populations which separated only a few generations ago, such as, for instance, Scots in Scotland and Scots in America, are likely to differ only slightly in gene frequencies.

Our object will usually be, however, to use the observed blood-group gene frequencies of two or more populations as indicators of the relationship between them. If gene frequencies of two populations are known only for one genetic system, then the more alike these frequencies are, the more closely related the populations are likely to be, and the more recently their ancestors are likely to have separated. However, the resemblance in such a case may be just a matter of chance. For instance, Eskimos and Scots, who are certainly not closely related, have rather similar ABO gene frequencies. Much of the early work on blood-group anthropology suffered from attempts to draw far-reaching conclusions from ABO frequencies alone. But if we examine frequencies not just for one but for several genetic systems and find them alike, then this is strong evidence of close relationship, and the more systems we use the stronger is the evidence. Moreover, if we examine a whole lot of populations for a whole lot of genetic systems, there are now available statistical methods of combining the differences for all the systems, so that we get a pretty reliable estimate of the closeness or distance of relationship between each pair of populations.

We shall now proceed to apply these principles to as many as possible of the peoples of the world, beginning with those of Africa.

It is almost certain that man evolved from his pre-human ancestors and emerged as a unique tool-making animal somewhere in tropical Africa and that we are therefore, in a sense, all of African origin. I once said this in the southern United States and was questioned afterwards by the press as to whether our original ancestors were 'coloured'; I caused some shock when I said they probably were!

It is likely that the human line separated from that of the great apes some ten million years ago, and that it is about one million years since the genus *Homo* arose. During these million years men have left Africa, for Asia in the first instance, and some evolution has taken place outside Africa. However, there has been sufficient interchange of genes between Africa and the outside world to ensure that the sole existing species, *Homo sapiens*, has indeed remained a single interbreeding species. For the most fully studied sets of characters, such as colour, form, and blood groups the indigenous inhabitants of tropical Africa differ genetically from other human populations to a greater degree than most of these differ from one another, so that it is tempting to imagine that they are closer to the earliest men than are any peoples outside Africa.

North Africa is inhabited by pale-skinned or caucasoid peoples whose ancestors entered from Asia only some 10000 years ago, and who have remained relatively unmixed because of the formidable barrier of the Sahara Desert, even if it was not always such a complete barrier as it is now. Similar folk at about the same time entered east Africa from the north but they have interbred in varying degree with the former dark-skinned inhabitants to give mixtures which are difficult to unravel. We shall deal with them later. We must first look at the really ancient dark-skinned peoples who, apart from recently immigrated colonies of Europeans, inhabit the rest of Africa.

Their physical appearance is well known; they have in varying degree dark skins, woolly hair, thick lips, and a number of other special facial characteristics. They also, however, almost universally show numerous particular features in the frequencies of the common blood groups and the occurrence of certain special antigens, which they do not share with the indigenous inhabitants of any of the other continents.

At once we meet the same difficulty as confronts us in all continents. We know far more about the distribution of the blood groups of the ABO system than about those of any other system. Yet ABO frequencies appear to be much more labile than those of most other systems, and so can be used with confidence only as indices of relatively recent population movements and mixings, perhaps up to one or two thousand years ago. Africans do, however, on the average have more *O* and *B* genes and less *A* genes than Europeans, and they have somewhat raised frequencies of two variants of the *A* gene; one is intermediate in properties between A_1 and A_2 and is known as A_{int}; the other is barely distinguishable from *O* and is known as A_x. Even in African populations these genes are quite rare, but they do help in sorting out the relations between some of the major classes of these populations.

The MNSs system is much more revealing. The frequencies of *M* and *N* are on the average about equal to one another and not greatly different from those found in Europe. The *S* gene however is considerably rarer than Europe, and those *S* genes which do exist are rather more evenly distributed in their linkage between *M* and *N* than in Europe. This is thought to be due to the relative freedom of the peoples of Africa as a whole, over a very long period, from the introduction of genes from outside the continent, so that by repeated crossing-over during tens of thousands of years the frequencies of the haplotypes or linked combinations *MS, Ms, NS*, and *Ns*, have reached a point nearer to linkage equilibrium than has occurred elsewhere, for instance in Europe. Africans have, however, a third allele, peculiar to themselves, of the *S* and *s* genes, known as S^u. It may produce a specific antigen, but no reagent has so far been found which will detect this, so that the gene can be detected only in the homozygous state, by the absence of reactions both for *S* and *s*. This gene has not been detected in all African populations and its varying frequency is of great interest in showing how populations are related. Another gene of the MNSs system, peculiar to Africans, is the Henshaw or *He* gene which gives rise to the Henshaw antigen. The gene is certainly closely linked to the genetic loci for MN and SsS^u. The gene, though never common, seems to be present in all the populations of tropical Africa, but its frequency and, more interestingly, its linkage relations, vary considerably. It is nearly always linked to *S* rather than *s* but in some populations it travels mainly with *M* and in others with *N*.

The Rhesus system also shows a number of features which are peculiar to Africans. Rh-negatives, and the *d* gene, have a substantial frequency; that of *d* varies around 20 per cent though it is much lower in a few populations. A frequency of 20 per cent is about half that found in Europe and since the frequency of the Rh negative phenotype is equal to the square of the frequency of *d*, this means that there are only about 4 per cent of Rh negatives as compared with 16 per cent in Europe. The problem of haemolytic disease of the newborn, which occurs almost solely in the children of *d*-negative women, is thus much less important than it is in European populations. The most striking characteristic of the Rh system in Africans is however the very high frequency of the *cDe* haplotype, which nearly everywhere exceeds 50 per cent, and is sometimes over 90 per cent, whereas in populations not of African ancestry it seldom exceeds 10 per cent. The cD^ue haplotype, with a frequency of a few per cent, is almost confined to Africans.

Another important African marker belonging to the Rh system is the V antigen. It is transmitted genetically as though a part of the Rh system, almost always along with one of the haplotypes *cDe*, cD^ue, *cde*. There is however some ambiguity as to its relation to the three major pairs of alleles, *C,c; D,d;* and *E,e,* and it is simplest to treat it provisionally as the product of a separate, closely linked, gene *V*. It is present in all African populations which have been tested for it, so that its presence in a population is a trustworthy mark of African ancestry. Though it is usually associated in individuals with the *cDe* haplotype, its geographical frequency distribution pattern within Africa is considerably different from that of *cDe*, and is, so far, not fully explained.

The *K* gene of the Kell system is rare in African populations but the Js^a gene of the same system is another well-defined African marker. It has been found absent only in one tropical African population. Its frequencies otherwise vary from 2 to 26 per cent but many more observations are needed to define its distribution pattern.

The Duffy system differentiates Africans even more sharply from non-Africans than does any of the other systems mentioned. Outside Africa the main genes of the system are Fy^a and Fy^b, these together accounting for about 99 per cent of the total, whereas in tropical Africa a third gene, Fy^4, accounts for over 90 per cent, apart from southern Africa where frequencies are somewhat lower. This is explained by the almost total resistance of Fy^4 homozygotes to infection by benign tertian malaria. The evolutionary process involved is further considered below (121–2). Because of the outstanding interest of Fy^4 as a marker gene as well as one participating in an important process of natural selection, it is most unfortunate that the anti-Fy^4 reagent remains extremely rare, so that the gene can be detected only in the homozygous state.

The P_1 gene, and the Jk^a gene of the Kidd system, are more abundant in Africa than elsewhere, but not sufficiently so to make them useful by themselves as markers.

Apart from the blood groups in the strict sense, some other hereditary blood characters are much commoner in Africa than outside the continent. Perhaps the most important of these is sickle-cell haemoglobin or Haemoglobin S, because of the resistance to infection with malignant tertian malaria possessed by heterozygotes. Though the gene for Haemoglobin S is found mainly in Africa it is not certain that it originated there, and certainly the frequency pattern seen in the continent, closely related as it is to the incidence of this form of malaria, is something which has developed rapidly, and would change again rapidly in response to such factors as malaria control or the lack of it. The relation between Fy^4 and Hb S is discussed on p. 122.

Beta-thalassaemia and glucose 6-phosphate dehydrogenase (G6PD) deficiency, both of which also confer resistance to malaria, are likewise fairly common in Africa but also in tropical and sub-tropical regions elsewhere. Certain variants of G6PD are peculiar to Africa. The distribution of these genetic factors has not been found so closely related to that of malaria as has been the case for Haemoglobin S. A variant of red-cell acid phosphatase, P^r, is peculiar to the peoples of southern Africa.

No other subdivision of the human race possesses anything approaching the wealth of known specific markers shown by Africans. This richness is almost certainly genuine but it must be admitted that several of the markers were discovered in the American Negro population which is so readily accessible to investigation, while similar markers in other non-caucasoid populations may have been missed.

The question now arises as to the relation of the distribution of these marker genes to the origins of the various peoples of Africa. Though to a non-African eye all Africans may look alike, there is probably more physical variety, and certainly more genetic variety, among the peoples of Africa than among those of Europe. However, for a preliminary examination, we may divide tropical Africans into 'Negroes' and 'the rest'. The peoples of West Africa are the most familiar representatives in Europe of the 'Negro' physical type, but the same broad type is found also over a vast area of central and southern Africa. 'The rest' consist largely of people of low stature, found in small isolated groups all over Africa but showing some degree of genetic resemblance to one another. It must, however, be admitted that there are also in Africa substantial populations of very great stature.

In order to focus our picture let us try to reconstruct some events taking place in Africa some 2000 years ago. The smelting of iron, and the cultivation of cereal crops, had already reached a narrow belt just south of the Sahara Desert, but these crops were unsuited to the great rain forest belt further to the south. The expansion of the enterprising Bantu-speaking peoples had already begun, associated with the adoption of iron smelting, and the cultivation of those few indigenous plants which were suited to forest belt agriculture.

Then there arrived in East Africa invaders from Indonesia, probably mainly in search of slaves, but bringing with them cultivated plants from the rainforests of their native lands. These were rapidly found to be suited, far better than any indigenous Africa plants, to cultivation in the African tropics and it was their adoption by the Bantu speakers that, together with iron smelting which they already practised, gave a major stimulus to their development. Vast hitherto thinly populated areas of central Africa became able to support much increased populations. The former inhabitants were largely driven into the less fertile and less accessible areas, and the Bantu speakers progressed steadily southwards.

Before we follow the more recent history of the Bantu speakers we must look at another closely related group of populations, that of the Negro peoples who inhabit the rain forest which forms the southern part of West Africa. They speak languages of the Congo–Kordofanian group to which the Bantu sub-group also belongs. They, like the Bantu speakers, have learned to cultivate the forest zone using plants introduced from south-east Asia. Their essential culture thus has an origin shared with the Bantu speakers whom they also resemble in their broad physical characteristics. The story of how the human population and their culture reached West Africa has not however been as fully worked out as has been the case with the movements of the Bantu speakers.

The blood groups of Bantu speakers and West Africans

Much of our research on the blood groups of West Africans was carried out at a relatively early date, before the discovery of several of the special African marker genes. Also, our knowledge of the relatively unmixed Bantu speakers is somewhat restricted, for we know them best in the regions where they have interbred with east or southern African peoples; these mixed populations must be considered later. As might be expected, however, there is a considerable resemblance between the known blood-group frequencies of the Bantu speakers of central Africa and of West Africans, especially those living near the south coast of West Africa. They have about 15 per cent of each of the *A* and *B* genes and 70 per cent of *O*. In West Africa *A* and *B* increase and *O* decreases in frequency as we go northward, so approaching the levels found in the peoples of the savannah zone, discussed later, but in central Africa there is no clear general pattern of variation.

The frequency of the *M* gene in both regions is near 50 per cent. In West Africa it decreases steadily northward into the savannah zone while in central Africa a large central area, not very fully surveyed, shows frequencies somewhat below 50 per cent. The frequencies of the *S* gene, linked to *M* or *N*, are about 10 per cent in West Africa, as compared to about 16 per cent in central Africa. The linked Henshaw gene is slightly less common in West Africa than in the centre, but, more interestingly, in West Africa it is mostly found combined in a single haplotype with *N* and *S* whereas in central Africa it goes rather more frequently with *MS* than with *NS*. Combinations with *Ms* and *Ns* are everywhere rare.

With regard to the Rh system, the frequency of the *d* gene is just above 20 per cent over most of West Africa, whereas it is a little below 20 in most of central Africa. The *cDe* haplotype varies around 60 per cent in West Africa whilst in central Africa it is everywhere over 60 per cent and increases southward to above 70 per cent.

It is difficult to interpret the distribution of the genes of the ABO system. It is possible that what we believe to be a rather labile distribution has been affected by recent marked changes in the environment of the populations concerned. As already mentioned, the gene distributions of the MNSs and Rh systems are thought to have a greater long-term stability. For *M, S, d,* and *cDe* the West Africans show distributions which are nearer to those of Caucasoids than do those of the central Africans, and in each case there is a further trend in the same direction in the peoples of the savannah. It is thus possible that the Negroes of West Africa have incorporated some caucasoid genes while the central Africans have not. A further possibility, to be discussed later, is that *all* Negroes have some caucasoid genes. The linkage relations of the Henshaw gene are likely to have been of very slow development by crossing over, and may point to some very ancient events which we do not yet understand.

Bantu speakers and Khoisan

We must now follow the migration of the Bantu speakers yet further southward, to the point where they met one of the largest and best defined groups of 'the rest', the San and the Khoikhoi, or Bushmen and Hottentots, now known jointly as the Khoisan. At about the same time, around 1500 AD, groups of European colonists began entering South Africa by sea and spreading a European type of agriculture northward. At first they met only the scattered Khoisan but soon they came up against the numerous and well-organized Bantu-speaking tribes. So much is generally accepted on the basis of historical, archaeological, and botanical records. But who are the Khoisan, and what is their relationship to the Bantu speakers and the rest of indigenous Africans?

When, some 35 years ago, I began to study blood-group anthropology it was well recognized that African 'Negroes' were a well defined 'race' and this was soon fully confirmed, in terms of the blood groups then known—in particular all the populations tested had high frequencies of the *cDe* haplotype, a high P_1 frequency and a low *S*. But the Khoisan were, to some extent rightly, regarded as a race apart. Many anthropologists regarded them as a separate 'major race' of similar status to Negroids, Mongoloids, and Caucasoids, and some even thought that they were more nearly related to Mongoloids than to Negroids.

Then, in 1953–5 their blood groups were for the first time examined by Dr A. Zoutendyk and his colleagues who showed that they represented a subtype of the general African population, with a particularly high frequency of the African *cDe* haplotype. Further work, largely by Jenkins, has filled in the details of their blood group picture, especially with regard to the various African marker genes. They have particularly high frequencies of the A_x allele of the ABO system, and of the Henshaw type of the MNSs system, as well as of the P^r allele of the red-cell acid phosphatase system. Js^a and Fy^4 on the other hand fall considerably below the usual African levels, and S^u is absent.

Following the pioneer work of Zoutendyk and his colleagues, the extremely detailed work of Dr Trefor Jenkins has shown that for a great many genetic systems the Khoikhoi and the San differ from the majority of Negro populations, that the San usually differ more from the Negroes than do the Khoikhoi, and that South African Negroes differ from nearly all other Negro populations not only in showing a quantitative divergence in their gene frequencies towards those of the Khoisan, but in tending to possess alleles, at various loci, which otherwise are almost entirely confined to the Khoisan.

While Jenkins's work tends to emphasize the differences between the Khoisan and the Negroes, the importance of Zoutendyk's work lay in showing that, while the Khoisan differed from the Negroes, the two population groups represented variations on a well-defined African pattern totally distinct from those of the other major races. Thus blood-group studies played a major part in disproving the claims of some anthropologists who regarded the Khoisan as belonging to a separate major race from the Negroes, and initiated studies now in progress which are beginning to show how a single African human stock became differentiated, and the various branches then interacted with one another, and with invaders from outside, who were only remotely of African origin. While however these processes must have taken tens, or even hundreds, of thousands of years to produce their effects, the present local pattern in South Africa itself is to be interpreted in terms of events in only the last thousand years. It appears likely that, less than a thousand years ago, the Khoisan were almost the sole occupants of the south of Africa, probably already differentiated into San hunters and gatherers and Khoikhoi herders. It further appears that the San genetic pattern is the nearer to that of the original Khoisan, and that of the Khoikhoi has been the more greatly modified by genetic drift and perhaps by natural selection and hybridization. The Bantu-speaking newcomers of recent centuries have largely displaced the Khoisan, but both their languages and their genes show that they have incorporated a great many Khoisan in their communities.

It has already been mentioned that the west African Negroes, though very similar genetically to the Bantu-speaking ones, are slightly more like Caucasoids, and that this may be due to actual incorporation of caucasoid genes from populations to the north.

If however we regard the Bantu speakers as 'standard' African, then the Khoisan are, in some respects, as we have seen, more African than the Africans. Another way of looking at this situation is to suggest that all Negroes are in varying degree African–Caucasoid hybrids. This is a hypothesis demanding careful consideration as genetic, anthropometric, linguistic, social, archaeological, and historical studies converge on trying to find out how the Bantu-speaking peoples originated.

East Africa

Before we can attempt to trace any further the possible former presence of people of Khoisan affinities in places remote from their present home in southern Africa, we must examine the complicated situation which exists in east Africa as a result of the arrival of people from outside Africa during the last ten thousand years or so.

Though man first achieved humanity in Africa, some of his physical evolution and much of his cultural development took place elsewhere, and especially in Asia. During palaeolithic times the mongoloid and caucasoid races appear to have evolved in Asia and Europe, but it was only towards the end of the palaeolithic that there was any massive re-entry into Africa. At this time, though not necessarily simultaneously, caucasoid peoples, who had developed a relatively advanced culture in south-west Asia, entered Africa in two main streams. One of these followed the relatively fertile north coastal strip or Maghrib, and their history is discussed below (pp. 48–50).

The other stream of late palaeolithic caucasoid invaders followed the irregular but almost continuous belt of high ground which runs down the east side of Africa, some hundreds of miles from the coast. Here the Nile Valley, the sea, and the highlands themselves have provided relatively easy highways for human movement between north and south; south of the desert belt there was never any sharp barrier to east-to-west movement.

Thus in mesolithic times we find almost identical cultures in the north African Maghrib, in Khartoum, and in Kenya, all apparently brought in by Caucasoids from south-west Asia. Also, to this day most of the inhabitants of east Africa down to as far south as Kenya speak languages of the northerly derived Afro-Asiatic group. There is, however, in east Africa

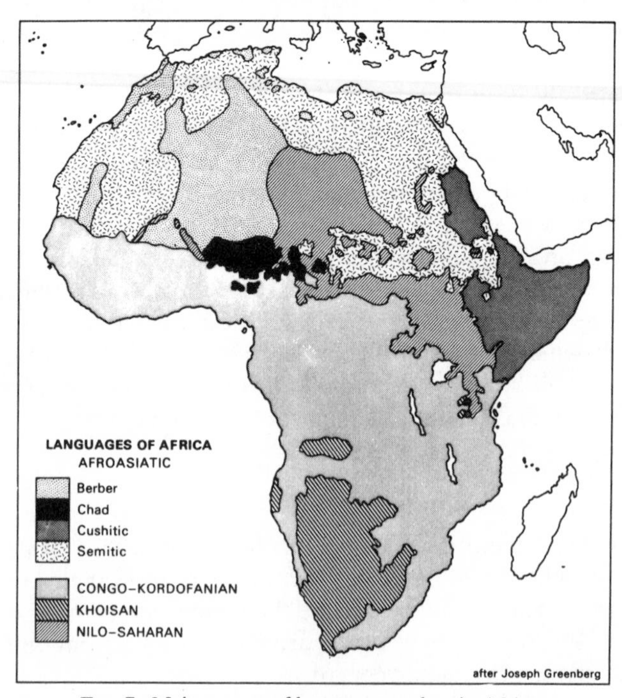

FIG. 7. Main groups of languages spoken in Africa.

nothing like the same clear-cut distinction between Caucasoids and Negroids that we find in north Africa, whether in terms of physique or of genetic markers. We cannot enter into the details of the complicated and specialized subject of morphological anthropology, but must explore the distribution of the blood groups, and particularly of the characteristic African marker genes in relation to the traditional physical, linguistic, and other cultural features which distinguish between populations.

In the northern half of the African continent a distinction can be made between the speakers of the Afro-Asiatic languages already mentioned and speakers of other languages, mainly Sudanic and Bantu (see Fig. 7).

Whereas farther west there is a fairly clear distinction physically and serologically, as well as linguistically, between Caucasoids and Negroids, this is not the case in east Africa where both archaeology and liguistics point to extensive and probably early incursions southward of caucasoid peoples, yet genetically the distinction is by no means clear, for most of the populations living south of latitude 10°N, whatever their language and culture, have a high frequency of the African marker genes, and the present blood-group distribution seems mainly to reflect population movement within the historical period. Perhaps therefore we have to suppose that, unlike the situation in the north, where a relatively numerous caucasoid people largely displaced the former inhabitants, the caucasoid invaders of east Africa, though they were able to introduce and impose their culture and language, became genetically merged with the more numerous indigenous Negroids.

There is, however, sometimes an apparent discrepancy between the indications of relationship yielded by different genetic systems or hereditary physical characters. For instance the cattle-herding Masai of Kenya have only moderately dark skins, and almost caucasoid facial features, but they have fuzzy hair and a thoroughly negroid blood-group picture. The frequency of the specific African marker genes is also greater in east and north-east Africa than might be expected from historical records and the physical appearance of the people. Even the Henshaw gene, thought to have originated in the proto-Khoisan, has moderately high frequencies in all east African populations tested for it. The *V* gene, universally present in Negroids, has a most peculiar distribution. It is clearly not primarily a Khoisan gene, but its highest frequencies, in a variety of ethnic groups, tend to follow the eastern highland belt of the continent, rising to three main peaks, one of 40 per cent in the Herero of Botswana, close neighbours of the San, one of 35 per cent in the click-speaking Sandawe of Tanzania, and one of 38 per cent in the Kunama of Eritrea. It is however at this northern end of the belt that there is the greatest concentration of high frequencies, and those found in the caucasoid Arabs of the Arabian peninsula are much higher than would be expected from their proportion of other African genes. Thus it may have originated as a mutation in north-east Africa, but long enough ago to spread to the whole of the continent.

We now return to the question of whether the Khoisan can be regarded as the descendants of a distinct branch of the Negroid race, once having a distribution comparable to that of the now commoner type of Negroes who inhabit most of the continent.

In southern Africa skeletons of San and Khoikhoi type go back far into prehistory, with no signs of Negro features.

As far north as Tanzania we find two tribes with clicks in their languages, the Sandawe and the Hadza. Linguistically and physically the Sandawe show the greater resemblance to the Khoisan, but with typically Negro blood groups. The Hadza on the other hand have remained hunters and gatherers like the San, and like them they have a very high frequency of *cDe*. We are here perhaps observing the effects of genetic drift and hybridization on populations originally closely related to the Khoisan.

Further north still, in central Africa, the main unusual peoples are the various tribes of pygmies. Physically their chief resemblance to the San is in their small stature; otherwise they appear to be more closely related to the Negroes, and it has indeed been suggested that they represent a major component in the ancestry of modern Negroes. They do, like the Khoisan, have a very high frequency of *cDe*. Even more surprisingly, the Northern Nilotes of the Sudan Republic (Dinka, Nuer, and Shilluk) who are very black, but are among the tallest people in the world, and are somewhat caucasoid looking, have very high frequencies of *cDe*. Perhaps then, we are catching glimpses of a very ancient African pre-Negro stock with varying physical characteristics, all with very high *cDe*, possibly the result of natural selection very long ago, and it may be that the typical Negroes of West Africa and the great central and south African belt of Bantu speakers with only about 60 per cent of *cDe* are, as already suggested, partly of Caucasoid descent.

Malaria, cattle, and Indonesians

As we have seen (p. 32), the gene for sickle-cell haemoglobin, or Haemoglobin S, is widely distributed in tropical Africa. Because heterozygotes for this gene have a high resistance to malignant tertian malaria (caused by the parasite *Plasmodium falciparum*) its frequency in different populations has, through natural selection, come to correspond with the distribution of the parasite. On this basis we can account for most of the details of the distribution of the gene in Africa. However, the southward distribution of the gene stops almost completely short at the Zambesi River. Another thing which stops at the same river is the distribution of the short-horned zebu, a type of cattle which originally came in from India and probably entered Africa through Egypt.

And sickling, malignant tertian malaria, and the short-horned zebu are all present in Madagascar.

To explain these facts we must now look at Madagascar and its people. The island of Madagascar, 1600 km long, 500 km wide, and only 260 km from the African continent, presents an extreme contrast to the latter in the nature of its human and animal inhabitants. It has been separated from Africa by the process of continental drift since some time in the Tertiary geological era, about fifty million years ago, before the evolution of the monkeys and the great apes, let alone man, so that the highest primates present (other than man) are the lemurs which are very common and of which the island possesses three quarters of all the known species, and very numerous individuals.

The Malagasy or Malgache, of whom there are about four million, all speak dialects of a single language which belongs to the Malayo-Polynesian group, with a few words of Arabic, Sanskrit, and Bantu origin; in the thirteenth century, after Arab colonization, the language was written locally in Arabic characters, but general literacy had to await its reduction to Roman characters by Christian missionaries in the nineteenth century. Thus the island has virtually no written history, and archaeology is inhibited by the local cult of the dead which persists in incongruous combination with Anglican episcopalianism.

The whole of the culture is essentially Malaysian and it is likely that the main peopling of the island took place from the South-East Asian archipelago early in the Christian era by colonists who also brought to Africa the food plants which formed the basis of the Bantu cultural revolution. These colonists have been supplemented by the extensive importation of African slaves, but island traditions speak of an early race of dwarfs, and it seems possible that the earliest inhabitants were related to the African Khoisan or Pygmies (or as some have supposed, to Asiatic Negritos, but this now seems unlikely). Malay, Negro, and intermediate physical types are found in all the tribes.

Blood-group studies have been few and patchy. The main feature which we can use for internal classification is the frequency of the *B* gene, which is an Asian rather than an African characteristic, especially since *B* usually exceeds *A* in frequency. The data are consistent with a mainly Malay origin for the Andriana or nobles and the Hova or commoners and a mainly African one for the Mainty, the former slave population. A single set of data shows the frequency of the *cDe* haplotype to be 37.5 per cent, below African levels but very much greater than in Malays, and suggesting that the population tested had about 50 per cent of African genes. The few other systems examined give results consistent with these.

Sickle-cell haemoglobin is present throughout the island, with a gene frequency of 4 per cent on the plateau and 15 per cent near the coasts. This is consistent with the fact that the coastal areas have been highly malarial and the plateau much less so. Both malignant tertian and benign tertian malarias are present.

Glucose 6-phosphate dehydrogenase deficiency, another character which confers resistance to malaria, is also present, but we do not know whether the gene responsible is of the African or the Malaysian variety (or both).

Much more information could be obtained on the origin of the various tribes if tests were done for the products of the various African marker genes. Moreover, if there was indeed an aboriginal pygmy population, this might have possessed high frequencies of some of the Khoisan marker genes already discussed.

As already mentioned, the cattle of Madagascar are mainly of the short-horned zebu type. It is likely that the southward migrations of the Bantu speakers down the east coast of Africa took place as a series of waves. One wave which reached south Africa brought neither the sickle-cell gene nor the short-horned zebu, whereas the arrival on the Mozambique coast of a subsequent one, bringing both, but stopping at the Zambezi river, coincided with the presence of Malaysian invaders. They were thus able to capture and take to Madagascar both slaves with their sickle-cell gene and their malignant tertian malaria parasite, and also their cattle.

North Africa

The part of Africa which lies north of the Sahara Desert forms a single natural geographical unit. Its history, as deduced from archaeological findings and from written records, is characterized by certain mass movements of populations, which one might expect to see reflected in the distribution of the genetic characters of their descendants.

As we have seen, man almost certainly originated in Africa and, perhaps about a million years ago, groups of men left Africa and their descendants spread to the greater part of Asia and Europe, evolving there into a number of major varieties of *Homo sapiens*, including those known as Mongoloid and Caucasoid.

Since man has remained a single interbreeding biological species there must have been repeated movements to and fro across the Isthmus of Suez during palaeolithic times, but the first major movement of human beings *into* Africa took place towards the end of that period, about ten thousand years ago. At this time Palestine, with the surrounding region, was the site of very rapid cultural development, and groups of people from this region, possessing an advanced stone-implement culture, but still only hunters and gatherers, not producers, of food entered Africa and, as we have seen, spread into two streams. One, which we have already tried to trace, followed the eastern highlands down to Kenya. The other spread along the narrow strip of more or less productive (and ultimately cultivable) land which stretches between the desert and the sea from Egypt to Morocco and is now known as the Maghrib.

The culture of these people, as shown by the implements they left behind, is known as Capsian, from the Latin name of Gafsa in Tunisia. To judge from their present-day presumed descendants, both streams of immigrants were moderately light skinned, and spoke languages of the Afro-Asiatic group. This group includes the Semitic languages (Hebrew, Arabic, etc.) and all the other languages at present spoken by the indigenous peoples of north Africa. The northern stream of immigrants, whom we are now trying to follow, may be regarded as consisting of proto-Berbers, ancestral to the 'pure' Berbers of the Atlas Mountains, and to the Berbers, now in varying degree Arabized in language and culture, of the rest of the Maghrib.

Rapid cultural advance continued in the Near East, with domestication of animals, cultivation of the soil, and all the other innovations of the neolithic stage of culture. Since the land could now support much larger numbers of people than was possible with the less intensive practices available to palaeolithic man, the spread of the neolithic revolution took the form of a self-propagating population explosion, with a movement not merely of a few people with ideas, but of large and progressively increasing numbers of persons. It was in this way that neolithic cultivators entered the Nile Valley and became established there in the second half of the fifth millennium BC. Archaeological investigation shows that at first they were a distinct group living side by side with the descendants of their hunter forerunners, but soon there was cultural and presumably genetic exchange. The farmers here found conditions uniquely favourable for agriculture, so that within about a thousand years of their arrival production was sufficient to support a highly sophisticated urban civilization. We shall not attempt to follow in detail the subsequent history of Egypt, but wars and changes of rulers must have brought many new elements from Asia into the population. Nevertheless, archaeology and the examination of the existing population show that through all this period the physical type of the Egyptian peasant has persisted with but little change.

We know little of the history of the Maghrib between its first settlement by the Capsian hunters and the establishment of settlements on the western parts of the coast of north Africa by Semitic-speaking Phoenicians about the eighth century BC, probably accompanied from the beginning by Jews bringing their unique monotheistic religion. Neither the Phoenicians nor their Roman and Vandal (Germanic) successors appear to have penetrated far inland, or to have modified greatly the genetic composition of the population as a whole. The Arab invasions of the sixth and seventh centuries AD brought about the conversion of most of the people of Egypt and the Maghrib to Islam, and the extensive adoption of the Arabic language. There must have been a substantial incorporation of invaders into the population but this has tended to be exaggerated because of the habit of Arabic-speaking Muslims, even if mainly of Berber stock, of claiming Arab ancestry.

The Muslims of the western Maghrib then went on to conquer most of the Iberian Peninsula. Many of the invaders were certainly of Berber origin. They were accompanied by considerable numbers of Jews. Finally, Christian armies reconquered the peninsula and vast numbers of refugees, both Muslim and Jewish, flooded into north Africa. The Jews have since remained largely distinct but the Muslims soon became incorporated into the indigenous population. The Muslim refugees must have been in part of Berber ancestry but they must have carried also a high proportion of genes of indigenous Iberian origin.

Perhaps, however, though it is little known other than to professional historians, the largest incursion of outside settlers into the Maghrib and its hinterland was the invasion of the Banu Hilal and other Bedouin tribes from Upper Egypt in the eleventh century AD. They were hardy but relatively uncivilized nomads, who largely completed the Arabization of the Maghrib, but at the same time caused the disintegration or most of its civilization. They, however, unlike the previous Arab invaders, took possession of much of the Sahara, especially its eastern parts. Hitherto only the Berber Touaregs had been able to live in its inhospitable wastes.

Against this background of archaeology and history we shall now examine the distribution of the blood groups in the peoples of north Africa, and try to see how far this distribution supports or supplements what we know from other sources. Since we shall, with very few exceptions, be dependent on the frequencies of the genes of three systems only, ABO, MN, and Rh, and since we know from historical sources that there was extensive Arab immigration into North Africa from the seventh century AD onwards, we shall first summarize briefly the distribution of these genes in the Near East whence the invaders came, and among the presumed purer Berbers of Morocco as giving an approximation to pre-Arab distribution.

Among the classic Arabs of the Arabian peninsula the frequency of the *O* gene is high, about 75 per cent, with about 13 per cent of *A* and 12 per cent of *B*. The sedentary Arabs of the Levant have far more *A* genes, up to 30 per cent, and correspondingly less *O*. The Berbers of the Atlas Mountains and the Touareg Berbers of the desert have *O* frequencies as high as the classic Arabs, with correspondingly low *A* and *B*.

The *M* gene, with frequencies a little over 50 per cent in large areas of Europe and Africa, averages nearly 70 per cent in the peninsular Arabs, and falls as low as 23 per cent in Moroccan Berbers. It is probably near 57 per cent in the Levantine Arabs.

The frequency of the *d* gene of the Rh system varies from about 42 per cent in northern Europe to 30 to 35 per cent in its Mediterranean parts. It is 26 per cent in the classic Arabs and 23 per cent in the Ait Haddidu Berbers of the Atlas Mountains in Morocco.

The custom of cohabitation of Arabs, and probably also of Berbers, with Negro slave concubines is long established and its effect in raising the *cDe* frequencies can be seen in numerous populations in the Near East and North Africa. The frequencies of the *ABO* and *MN* genes in Negro populations likely to have provided slaves do not differ widely from the average frequencies found in the peoples of north Africa, and therefore a moderate amount of Negro admixture will not seriously affect the use of these systems in tracing relationships between the various caucasoid populations. The opposite is true of the Rh system, for the caucasoid peoples of the Mediterranean area do not differ widely among themselves in the frequencies of their Rh groups. On the other hand, all Negro populations have over 50 per cent of the *cDe* haplotype, whereas in what may reasonably be taken as unmixed caucasoid populations it seldom exceeds 5 per cent. The frequency of this haplotype is thus a sensitive indicator of Negro admixture in this region as elsewhere. Tests for other African marker genes have only in a few cases been done on north African populations.

The frequencies of the ABO and MN groups thus provide indications of population movements and mixing in an east–west direction in the long narrow Maghrib, while the Rh groups tell us about south to north movements across the desert.

Egypt

As we have seen, the originators of the Egyptian civilization probably came in from somewhere near Palestine in neolithic times. However, since Egypt is so near the Isthmus of Suez, the sole land gate between Africa and the rest of the world, it would not be surprising to find a considerable mixture of peoples in modern Egypt. Nevertheless, as we have seen, the Egyptian physical type has been remarkably persistent. Unfortunately, few data exist for the Egyptian population for any system except ABO. For this system frequencies show a considerable uniformity over the Nile Delta and the Nile Valley as far south as Asyût. In the whole of this area there are only relatively slight variations from 25 per cent of *A* genes and 20 per cent of *B*. The profession of Coptic Christianity has sometimes been regarded as a mark of ancient Egyptian descent, but Copts and Muslims do not differ significantly in their ABO frequencies. The remarkable feature of these frequencies is the consistently high level of *B*, in marked contrast to the low values found not only in the Arabs of the Arabian Peninsula but in all other neighbouring populations as well.

The *M* gene frequency is fairly uniform at about 53 per cent throughout lower Egypt, and there is little difference in this respect between Copts and Muslims. This frequency is, however, much less than is found in Arabia, and somewhat lower than in the Levant.

Data for the Rh system show considerable scatter, but it is clear that the present population of Lower Egypt is essentially caucasoid though it incorporates at least 20 per cent of negroid genes.

Thus the Egyptians, though they differ widely from the Arabs of Arabia, show a certain parallelism with them, in being a well-defined nearly homogeneous population which has incorporated a fair proportion of African genes.

It is however difficult to account for the special feature of the Egyptian population—the high *A* frequency and even more so the high *B*. The latter in particular cannot be explained by immigration of any stock now known to inhabit adjacent countries. Perhaps however they are due to immigration in the long distant past or, if there has been a secular change since the main immigrant peopling of Egypt took place, it must have been caused by natural selection, for the country has for a long time supported a population so large that no significant genetic drift can have occurred. There is a wide field for speculation as well as research into the possible agent of such a presumed selection. Perhaps, indeed, the very density of the population has been an important selective agent.

The Nubians who live south of the First Cataract of the Nile have about 20 per cent of *A* genes, but only 10 per cent of *B*, and so differ widely from the people of Lower Egypt. Unfortunately we know nothing at the moment of their other blood groups.

The Maghrib

The peoples of almost the whole of the Maghrib from Libya to Algeria, show a remarkable approach to uniformity in the frequencies of their ABO and MN groups, with *A* and *B* gene frequencies just above 20 and 10 per cent respectively, and *M* about 50 per cent. Only a relatively few isolated populations show any substantial variation from these frequencies. Frequencies of the Rh blood groups, and particularly of the negroid *cDe* haplotype are much more variable, indicating considerable variations in the degree to which the populations concerned have incorporated African genes.

The 'Arab' population of Libya, with only 3 per cent of *cDe*, can have incorporated few Negro genes, while that of Tunisia with 15 per cent has taken in many more. Djerba, said to be the classical Land of the Lotus-eaters, is an island off the Tunisian coast which has long been a place of refuge for religious minorities, both Jews and Arab sects. Perhaps because they have undergone genetic drift there is little correspondence between the blood-group frequencies of these sects and their supposed origins, 'Berber' on the one hand or 'Arab' on the other, but some of them have remarkably high *A* frequencies, reaching 29 per cent in the Wahabi.

In Algeria the ethnic distinction commonly made is between the 'Arabs' and the supposedly more strictly indigenous Kabyles, but both have typical Maghrib ABO frequencies of just above 20 per cent of *A* and 10 per cent of *B*. However, the relatively unmixed Berbers of the north-west corner of Algeria have only 18 per cent of *A* and just under 10 per cent of *B* genes, with only 3 per cent of *cDe*. Even the cosmopolitan Arabs of nearby Oran have only 6 per cent of *cDe* but the Kabyles of Tizi Ouzou, east of Algiers, have 28 per cent of this haplotype.

A remarkable isolate is that of the Flittas who live at Zemmora, south-east of Oran, and have a long history of fierce resistance to successive rulers of Algeria. They are probably of Berber origin, perhaps with some admixture at an early period, but they appear to have constituted a strict isolate for many hundreds of years. Their ABO frequencies are unique, with 18 per cent of A_2 genes, the highest frequency known except in the Lapps of northern Scandinavia. The total *A* gene frequency reaches the high level of 30 per cent, and *M* at 57 per cent is also well above the general north Africa level. The presence of 26 per cent of *cDe* shows considerable Negro admixture, presumably long ago. The high A_2 frequency remains a mystery—it is presumably the result of genetic drift or natural selection, but why should only three known populations, the Lapps, the Flittas, and the Nagas of Burma (with 17 per cent of A_2) have evolved in this way, when the great majority of isolates in Europe and the Mediterranean area, including the Arabs of Arabia and the Berbers of the Atlas Mountains, have nigh *O* and low total *A* frequencies. It would be of great interest to carry out A_1A_2 subgrouping on the other high *A* peoples of north Africa.

Morocco

In contrast to all the rest of north Africa Morocco is covered by a close network of ABO determinations, though only a few surveys have included tests for other blood-group systems. As might be expected in a mountainous country with hundreds of separate Arab and Berber tribes and sub-tribes, there are considerable variations in blood-group frequencies, from the typical Maghrib Arab values of a little over 20 per cent of *A* and 10 per cent of *B* genes, in the coastal areas, to the extreme example of the isolated Ait Haddidu tribe of Berbers in the Great Atlas Mountains with only 6 and 4 per cent of *A* and *B* genes and 89 per cent of *O*, the highest frequency found among the many isolates of Europe and the Mediterranean area. This tribe is extreme also in other ways, having only 27.5 per cent of *M* genes (and 72.5 per cent of *N*). Both the high *O* and the high *N* are typical of the Berbers and similar but less extreme ABO and MN frequencies have been found in several neighbouring tribes. Unfortunately, outside the Great Atlas range we have little but ABO data, but in general throughout Morocco the Berbers have more *O* and less *A* and *B* than the Arabs. The Ait Haddidu also have a low frequency of the *d* gene, of 23 per cent, or 5.5 per cent of Rh-negative individuals. They have, however, 27 per cent of the *cDe* haplotype and 10.5 per cent of the gene for V, another African marker. They have also a rather high P_1 frequency and a low Fy^a, both typically African but, on the other hand, hardly any of the African Js^a gene. Elsewhere in Morocco *M* gene frequencies vary from the typical Berber frequency of 43 per cent to the more characteristically Arab 55 per cent.

Considering now the Maghrib as a whole, the blood-group distribution is seen to follow a simpler pattern than might have been expected from the complex history of its peoples. The Egyptians stand out as a uniform and unique population, showing little similarity to any neighbouring peoples, and certainly not to the Arabs of Arabia. Their *A* frequencies are not unlike those of the sedentary Arabs of Palestine and Trans-Jordan, and perhaps reflect their origins in the early neolithic cultivators of this area some 7000 years ago. Their *B* frequencies, however, are higher than those of any other adequately tested population nearer than Iran (apart from two small sets of Turks in southern Anatolia).

In the Maghrib itself, as we have seen, ABO frequencies are remarkably near uniformity over the whole 6000 km strip, apart from the Atlas mountains. MN frequencies are probably also nearly uniform, but data are far fewer than for ABO. In Morocco, and to a lesser extent in Tunisia and Algeria, those tribes which have maintained Berber customs and language tend to have higher *O* frequencies than the rest, and it may be that part of the higher *A* and *B* frequencies found in the majority of the peoples of the Maghrib are the results of Arab admixture. If so, however, the Arabs concerned were not mainly those of the Arabian peninsula, with their high O frequency. Moreover, if such incorporation had substantially altered the genetic composition of these peoples one would expect it to be very variable in its effects.The high degree of uniformity found suggests that the frequencies observed are, on the contrary, the legacy of a single uniform population entering north Africa and becoming strung out along its length. It is therefore suggested that the original 'caucasoid' peopling of the Maghrib in late palaeolithic times took place in two waves, first the ancestors of the desert Touaregs and the Atlas Mountain Berbers, with very high frequencies of *O* and very low *M* and *d*, who went to the extreme west, perhaps driven before the second wave, the main Maghrib population, with about 20 per cent of *A*, 10 per cent of *B*, 50 per cent of *M*, and 30 per cent of *d*. On this interpretation, the numbers of genes contributed by Phoenicians, Romans, and Vandals, and by the Arabs of the first invasion, were very small, despite their great cultural effects. On the other hand, as discussed below, the later Hilalite Arab invaders were of similar genetic constitution to the Arabs of the peninsula.

The peoples of the desert

We have already looked in some detail at the fully negroid peoples of tropical Africa and at the caucasoid ones of the fertile strip between the desert and the sea. We shall now look at the peoples of the intervening zone, those of the desert itself, and those of the belt between the desert and the equatorial forest region. Our approaches to these two zones will differ from one another. In the case of the desert peoples we know quite a lot about a fair number of isolated populations, and we have reason to believe that their entry into the desert, if not their origins by hybridization, has mostly taken place since the beginning of the Christian era, so that we shall try to compare them in some detail with the peoples of neighbouring lands. The peoples of the zone immediately south of the desert, on the other hand, appear to have originated longer ago, so that a broader approach is needed. Also, in the western part of this zone data are very few.

Our main genetic information on the desert populations comes from the work of Professor J. Ruffié and his colleagues. Most of the peoples concerned live in the Algerian zone of the desert but it must be borne in mind that, for historico-political reasons, Algeria includes the greater part of the desert hinterland of Morocco, Tunisia, and even part of Libya.

In the Libyan Fezzan ABO frequencies resemble those of the Maghrib but the frequency of *cDe*, 21 per cent, shows the presence of a considerable Negro component.

In the northern part of the desert, and based on Metlili (Algeria) live the Chaambas, who are regarded as the descendants of the Arabs of the Hilalite invasions. Their *A* and *B* frequencies of 18 and 6 per cent are distinctly lower than those of the peoples of the Maghrib, and nearer to those recorded in Arabia. Their *cDe* frequency of 20 per cent, however, shows them (like the Arabians, indeed) to have incorporated a considerable amount of African ancestry.

Closely similar ABO frequencies occur in the widely spaced Arab populations of Beni Ounif (described as Arabo-Berbers) of the Ahaggar (Chaambas) and of Saoura and Tidikelt, and all these are thus likely to be largely of Arabian origin. The Arabs of the Ahaggar, Saoura, and Tidikelt have respectively 41, 30, and 50 per cent of *cDe*, the last nearly approaching the Negro level. They have respectively 56, 60, and 50 per cent of *M* genes, the first two somewhat resembling those found in Arabia itself.

Associated with many of the Arab populations are groups of the more negroid-looking Haratines, sedentary cultivators of the oases, always with higher frequencies of *B* and of *cDe*. They are generally regarded as the descendants of freed slaves but this view is not universally accepted.

Throughout the southern and western parts of Algeria, and extending southward into the Aïr highlands of Niger are found the Berber Touareg, the blue-veiled folk, who, as already mentioned, were masters of this part of the desert long before the arrival of the Arabs. They show the low frequencies of both *A* and *B* which typify the 'pure' Berber populations. Similar frequencies are found in the Berber Chleuh of Saoura yet, further west still, near the western tip of southern Algeria, the Reguibat claim to be Arabs, and their *A* and *B* frequencies of 22 and 11 per cent are more Arab than Berber. Yet further west again in the Hamada of the former Spanish West Africa most of the Moorish tribes have low *A* and *B* frequencies like the Berbers.

Thus, throughout the desert zone, most populations, despite Negro admixture, can be distinguished as either Arab or Berber, and it is clear that Arab tribes, originating in Arabia itself as late as the eleventh century, did approach or even reach the westermost extremity of the desert.

North-east Africa

The peoples of the region constituted by the Republic of the Sudan, Ethiopia, Somalia, and the small French territory of the Afars and Issas are well defined anthropologically but difficult to classify genetically because they possess confusing combinations of negroid and caucasoid characters.

They all speak Afro-Asiatic languages, with the exception of the inhabitants of the south-eastern Sudan, an area for which, because of political unrest, we possess no blood-group data at all. Despite the linguistic characteristics the region is one where some very high frequencies of African marker genes are reached, and history and genetic data agree in showing it to be one where for thousands of years there has been confrontation and mixing between caucasoid and negroid peoples.

The main feature of ABO distribution in the region as a whole is the marked westward diminution in the frequency of *O* from the high levels found on the east coast together with a rise of *B* and, in certain areas, especially Ethiopia, a rise of *A*. The high *O* and low *B*, especially in the Beja of the eastern Sudan and the Somalis, presumably indicate Arabian admixture in the population concerned, as might be expected from what we know of their history.

This, however, is not the whole explanation, for both the Beja and the

Somalis have relatively low *M* frequencies near 50 per cent, as compared with over 70 per cent in all peninsular Arabian populations. On the other hand, nearly all the peoples of Eritrea on the Red Sea, and of the rest of Ethiopia, have high *M* frequencies suggestive of Arabian admixture, though only those near the coast have high *O* frequencies as in Arabia.

Since we think that ABO frequencies are more labile than MN ones it may be that the discrepancy between the two systems is due to the MN distribution in some way (we do not yet know how) indicating earlier events in population development than that of ABO.

It is however of interest to look at Rh and especially *cDe* distribution as an index of negroid contributions to these populations. The frequency of *cDe* is 43 per cent in the Beja of the Sudan as compared with average frequencies of about 17 per cent in Arabia and 20 per cent in Egypt. To the south, in Ethiopia, including Eritrea, it has a frequency nearly everywhere exceeding 50 per cent. It is almost precisely 50 per cent in the Amhara, the most fully studied population group in Ethiopia, who, on the basis of all African marker genes combined, appear to be about 50 per cent negroid and 50 per cent caucasoid.

We have already looked at the Northern Nilotes of the Sudan with their extremely high *cDe* frequency. It may be that the unexpectedly high frequencies of this haplotype in the peoples of north-east Africa, who on other grounds appear to have a large caucasoid component, is due to their having received their *cDe* from ancestors related to the Northern Nilotes.

Between the desert and the forest

Stretching westward from the Sudan border almost to the Atlantic is a broad belt of savannah country with rather sparse rainfall and scattered trees. It received the neolithic cultural revolution at an early date, presumably from Egypt, and supported a cereal-based agriculture while the forest zone was still inhabited by hunting-and-gathering populations.

We know little of the history of this region, and very little of the blood groups of its inhabitants who are essentially negroid. Our data come almost entirely from the western part of the savannah area, for we know virtually nothing of blood-group distribution in Niger and Chad, apart from the desert peoples of their northern regions who have already been described. From the very scanty data available from the rest of the savannah all the peoples of the area appear, like those of the northern part of the forest zone, to have rather high frequencies of the *B* gene, between 15 and 20 per cent, and similar frequencies of *A*, with *O* just under 70 per cent. With remarkable regularity, as we enter the forest belt inhabited by typical Negroes, *A* and *B* both fall below 15 per cent and *O* rises above 70. Near the west coast, apart from a slight fall in *B*, frequencies differ little from those found inland. The peoples of Senegal, however, have high *O* frequencies like those of the desert belt.

The MN frequencies show features of considerable interest. We have already seen that the Berbers of Morocco have some of the highest *N* frequencies known. The high-*N* region continues down the west coast of Africa and several of the peoples of Senegal have *M* frequencies below 40 per cent. Inland in the savannah belt *M* frequencies are mostly between 40 and 50 per cent.

Despite these unusual MN frequencies, the Rh groups of these populations are typically African with *cDe* frequencies mostly exceeding 50 per cent. However frequencies in the savannah belt tend to be lower than nearer the south coast of West Africa.

It is likely that all the features in which the people of the savannah differ from those of the forest come from somewhere in the north—the high *N*, the rather high *A* and *B*, the slightly lowered *cDe*. The last could have come from anywhere in the north, the high *A* and *B*, along with the knowledge of cereal cultivation, ultimately from Egypt. But the high *N* must have come from the Berbers, now almost confined to Morocco, or be the result of natural selection under desert conditions, which is made unlikely by the very high *M* of the Arabs of the Arabian desert. It is likely that more blood-group data from this little-studied region would help to solve the intriguing historical and anthropological questions which it poses.

4. Asia

It was probably rather over a million years ago that man entered Asia from Africa; bones of the early human species, *Homo erectus*, have been found in China, as well as in Java which could have been reached only through Asia. Moreover, to reach Europe, which he probably very soon did, he must have passed through south-west Asia. It probably was in Asia that Eurasian man, by now of the modern *Homo sapiens* species, diverged from African man, and then became differentiated into the caucasoid and mongoloid types, but archaeology tells us little as to where, when, or how this occurred. Another differentiation which probably took place in Asia is that of the Australoids, perhaps from a common type before the separation of the Mongoloids.

Caucasoids and Mongoloids show some fairly definite blood-group differences but the two 'races' resemble one another much more closely than either of them does the Negroids. The differences are largely quantitative ones of gene frequencies rather than the qualitative presence or absence of particular genes.

Mongoloids have, in general, more B, more M, less S, less P_1, and more Fy^a than Caucasoids. They have a much lower frequency of the d gene, with more CDe and cDe. There are however some more nearly absolute differences. Unmixed Mongoloids probably lack the K gene completely; and they have a low but consistent frequency of Di^a which is virtually absent in Caucasoids.

The difference between the two 'races' appears rather sharp as we cross the mountains on the northern boundary of the Indian sub-continent. The passage from Indian to Burmese is somewhat more gradual, probably because contact here has been present for a very long time and some mixing has taken place, whereas the Mongoloids north of the mountains were probably fully differentiated in the Far East before the retreat of the ice allowed them to enter Tibet.

However, before we consider further these two major races, we must look at some small populations which may represent an earlier major race, akin to the Australian aborigines, and perhaps also to the peoples of the interior of New Guinea. These are the so-called Veddoid tribes found in the Nilgiri Hills of south-west India.

As we shall see, man has probably been in Australia for over thirty thousand years, so if the 'Veddoids' are indeed descended from these same people who originally went to Australia, they must have been in India for a very long time, and as their numbers are now small, they must have undergone considerable genetic drift, so that their present gene frequencies may differ a great deal from those of their cousins who went ultimately to Australia. The term 'Veddoid' is a complete misnomer. They do not resemble the Veddas of Sri Lanka (whom we shall consider later) either physically or in their blood groups. However, these tribes, the Paniyans, Chenchu, and Kurumbas, resemble one another and also the Australians in having high frequencies of the *A* gene and of the *CDE* haplotype. They also, unlike the Australians (but like many Africans), have the Haemoglobin S gene, and a rather high frequency of *M*. In the first three of these characteristics they also resemble the Padhu or Bathgamuwa of eastern Sri Lanka and it is these, rather than the Veddas, who are probably the true aborigines of that island. There are a number of so-called aboriginal populations also in Malaysia; we shall consider them later, as it is by no means certain that they were there before the more typical Mongoloids.

The Mongoloids are the most numerous of the three major races of mankind, and China, in the centre of the mongoloid area, has the largest population of any country in the world, so the Chinese must be taken as the typical Mongoloids (always remembering, however that Mongolia itself is in the extreme north of China). Unfortunately, because of recent political events, the blood groups of the Chinese have not been studied as fully as those of some neighbouring populations, notably the Japanese, and most of our data on the Chinese are derived from studies on emigrant populations. However, with the very recent revival of academic science in China it may be expected that the lack of data will soon be made good. Ninety-four per cent of the population of China claim to belong to the Han ethnic group. Genetically as well as culturally the Han people, despite their vast numbers and the very large area which they occupy, are a highly uniform one, and seem to have remained so because of the relative flatness of the country and hence the relative ease of internal communications. They thus form a most convenient population with which to compare populations on all sides of China, and minorities within the country itself. It is of course likely that even among the Han, as has for instance happened among the English, more detailed study than has so far been possible will reveal considerable minor heterogeneities. However, as far as can be judged from existing knowledge, the frequencies of the *A* and of the *B* genes each differ little from 20 per cent, and that of *M* is everywhere near 50 per cent, which is below that of most other populations in eastern Asia. In the Rh system most of the haplotypes (around 70 per cent) are *CDe*, and the rest mostly *cDE*, the latter, as in nearly all other parts of Asia and Europe, increasing steadily northward. The *d* (Rh-negative) gene is almost entirely absent.

To the south of China, extending from the Vietnamese border to the tip of the Malay peninsula, is a land of great river valleys running from north to south and separated by long mountain chains which effectively prevent movement across them and mixing with populations on either side. This has led to the region being occupied by a very complex mosaic of ethnic groups. Through this area there must have passed very long ago *Homo erectus* on his way to Java, then perhaps some forty thousand years ago the ancestors of the Australian aborigines (of whom, as we have seen, traces remain in southern India, off the main route). The ancestors of the inland peoples of New Guinea must also have passed this way but none of their descendants have been identified in continental Asia. Another quite ancient and physically distinct group are the Negritos of whom some remain in inland Malaysia as well as on various islands. The aboriginal Senoi of Malaya are also probably an ancient population. Then came the Indonesians, and finally the Mongoloids in a narrower sense, represented by the Mons and the Khmers, the Tibeto-Burmans and the Thais, all of whom probably entered the region before 1000 BC. It is rather ironical that, while the Khmers, or perhaps a small group of them, have recently become known as one of the greatest culture-destroying peoples of all time, they and the Mons are best known historically for the Indian type of high civilization, with its wealth of sculpture and painting, which they created in the lowlands of southern Burma, Thailand, and Cambodia. Even the Vietnamese, one of the last major ethnic groups to enter the area, arrived from south China as long ago as 200 BC.

Despite the extreme ethnic complexity of the peoples of this area, which is to some extent shown in the wealth of data on their ABO frequencies, there is an underlying unity, differing from that of China. They have considerably higher frequencies of the *M* gene and many, perhaps nearly all, of them have substantial frequencies of Haemoglobin E which is peculiar to this part of the world. By analogy with African (and south Indian) Haemoglobin S we may suspect that its presence confers resistance to some form of infection, possibly malaria itself, as in Africa. Indeed it has been suggested that the failure of the ethnic Vietnamese to establish themselves in the malarious highlands of Vietnam is related to their low frequency of Haemoglobin E, and consequent lack of resistance to the infection.

The Negritos of central Malaya, and of the Andaman Islands, bear a certain physical resemblance to the Pygmies of central Africa and it has long been argued whether or not there is any genetic relationship to the latter. The Malayan Negritos do have a rather high frequency (27 per cent) of the *cDe* haplotype, but the Andaman Islanders, who are probably less mixed, have no *cDe* at all, only *CDe* and *cDE*. A loss of *cDe* could be due to genetic drift but the high *CDe* suggests a relationship to the Melanesians to the east, rather than to Africans.

The 'aboriginal' Senoi of Malaya have the *O* and *B* genes but hardly any *A*. In this they resemble on the one hand the Tarons and the Htalu of the Sino-Burmese border, and on the other the Veddas of Sri Lanka. There is almost certainly a genetic relationship to the latter, for the Veddas are almost the only population outside south-east Asia known to possess Haemoglobin E.

Perhaps related to these high-B peoples are the Miao minorities in China itself. Both the Chwan Kiao and the Pa Miao have about 30 per cent of *B* genes, and the former have only 13 per cent of *A* genes.

Tibet is politically part of China but the Tibetans, on the basis of rather scanty data, differ considerably from the Chinese proper. In particular they have rather high frequencies of the northern Rh haplotype *cDE*; and a more typical Asiatic frequency of about 63 per cent of the *M* gene.

The western part of Taiwan is inhabited by a rather mixed population of Chinese resulting from a succession of invasions over the last 1500 years, but in the central and eastern more mountainous region, there are a number of 'aboriginal' populations whose blood groups resemble rather those of the peoples of south-eastern Asia, especially in their high frequencies of *M* and of *CDe*.

The Koreans differ little in their blood-group frequencies from the northern Chinese, but the Japanese, whose ancestors almost certainly passed through Korea to reach the Japanese islands, differ markedly from both. We shall however look at them again after studying the peoples of Japan.

When the main body of Japanese arrived in their present island home they found the islands inhabited by the ancestors of the present Ainu, now confined to the northern islands of Hokkaido and Sakhalin (the latter now in Soviet hands). Both physically and serologically the Ainu differ considerably from all the other peoples of Asia. In particular they are markedly hairy, the men having very full beards. It has even been suggested that they are an ancient offshoot of the Caucasoids, but their blood groups, despite the differences from other Mongoloids, do not on the whole support this theory. Perhaps, however, their ancestry goes back to a time when the Mongoloids and the Caucasoids were not as fully differentiated as they now are. Their ABO groups, which in any case are probably not an ancient feature, do not differ widely from those of neighbouring peoples. The same is true of most of their other blood-group frequencies, but their Rh and MNSs picture is in several respects unique. In the Rh system they have the totally unique frequency of 16 per cent of *cdE*, together with a small frequency of *V*, combined, however, not with *ce* as in Africa (as *cde*, $cD^{u}e$, *cDe*), but with *CDe*. There is no question of this haplotype being derived from Africa; the *V* gene, whatever its precise nature, is probably the result of a separate mutation. They have also a frequency of the *N* gene, unusually high for Asia, of 59 per cent, which includes 21 per cent of *NS*, one of the highest frequencies known of this haplotype.

Blood-group anthropological studies began at an early date in Japan, and the Japanese are among the most fully studied populations in the world, at least with regard to the ABO system. Together with a moderately high frequency of *B* they have, by a small margin, some of the highest *A* frequencies in eastern Asia. The wealth of data on the ABO frequencies of the Japanese has enabled highly detailed distribution maps to be drawn, comparable only to those available for the United Kingdom and a few other European countries.

Of greater general interest however, are their Rh and MNSs groups. In the Rh system they have 4.65 per cent of the *cdE* haplotype, one of the highest frequencies known other than in the Ainu, and they have a rather high *NS* frequency for an Asiatic population, 3 per cent, as compared for instance with 0.5 per cent in the Chinese. This in fact, implies a state of linkage disequilibrian; the parallel situation in the Polynesians is more fully discussed on p. 106. The above two resemblances to the Ainu convey a strong suggestion that the Japanese population carries a component of the order of 20 per cent of Ainu genes. It would therefore be expected that if tests on adequate numbers were done they would show also a small but informative frequency of the *V* gene, the other important Ainu marker.

If we now look again at the Koreans (especially in comparison with the neighbouring Chinese) we see that they too have rather high frequencies of *NS* and *cdE*, suggesting that they too carry a small proto-Ainu component, though a smaller one than the Japanese.

Soviet Asia and Mongolia

It is only very recently that work by Russian investigators has begun to supply anything beyond a scattering of ABO data on the populations of Siberia. These populations have an intrinsic interest of their own, but they are also important in supplying evidence regarding the original peopling of the American continent. During the last ice age Siberia, though very cold, was not largely ice-covered, as were most of the northern parts of Europe and America. The world sea level was much lower than it now is and, over a period of tens of thousands of years, Bering Strait and a very extensive surrounding area now covered by sea were dry land. There is still considerable controversy as to when man first entered America, but it was certainly over this tract of land, and perhaps about thirty thousand years ago, that populations of mongoloid physical type moved into north-eastern Siberia and thence into America.

It must have remained very cold in north-eastern Siberia for tens of thousands of years, but there must have been considerable fluctuations in the coldness, and it is likely that man's progress towards and across this region was largely controlled by such fluctuations.

When blood-group surveys were first extended to the Amerindians and to the peoples of eastern Asia, the differences in blood-group frequencies on the two sides of the Pacific were at once seen as an anomaly. Modern Amerindian populations have very few *B* genes or none at all and those of Central and South America also have few or no *A* genes. It is indeed probable that *B* was totally absent from Amerindian populations before the advent of Europeans, and that *A* was similarly absent from Mexico southwards. The *M* gene and the *cDE* haplotype have some of their highest known frequencies in the Amerindians. On the other hand the Chinese and the Japanese, who were the first far eastern peoples to be studied, both have moderately high *A* and *B* frequencies; they have approximately equal frequencies of *M* and *N*, as in Europe, and *cDE* has only a moderate frequency.

It was soon realized, however, that the peoples of Siberia were likely to be more closely related to the Amerindians than were the Chinese and Japanese, and that they might have blood-group frequencies very different from the latter.

Thus recent work on the peoples of Siberia has had the double interest already indicated. However, in proportion to the large populations involved, and to their great physical variety, the data at present available must still be regarded as very scanty, and no Asiatic populations have been found showing all the genetic characteristics of the American Indians. The *B* gene is everywhere present in Siberia but it falls to 8 per cent in the Chukchi who are the main inhabitants of the extreme east; the Yukagir of the frozen *taiga* to the west of them have only 7 per cent. The *A* gene also has moderately low frequencies, between 15 and 20 per cent, in much of northern Siberia. The *M* gene has high frequencies, similar to those of the American Indians, and much higher than those of the Chinese and Japanese, over most of north-eastern Siberia, and *cDE* has moderately high frequencies, though by no means as high as those of the American Indians. Farther west, but extending for thousands of kilometres along the north Siberian coast, are populations with a very high frequency of *N*, but these are largely, perhaps mainly, of western origin and will be discussed later (p. 72).

The Himalayas

Another area where man's entry must have been limited by the climatic factor was the Himalayas, for these remained glaciated and inaccessible until the temperature had somewhat ameliorated from the minimum which occurred some twenty thousand years ago. Thus the present inhabitants of the higher parts of Nepal and Bhutan appear to have followed the recession of the ice from western China and Tibet. They are thus of markedly mongoloid type, and fairly sharply differentiated from the caucasoid peoples of the lower and more southerly parts of these countries; there was nevertheless probably some mixing in an intermediate zone. The peoples of the highest zone of Nepal, the Sherpas, and most of the peoples of Bhutan, are physically the most like the Tibetans and have high frequencies of the *B* gene. Most of the peoples of the intermediate zone, notably the Gurung and Magar of Nepal, nevertheless show some northern features, notably the presence of the Mongoloid Di^a gene, and high frequencies of *cDE*.

The caucasoid peoples

As we have seen, the Caucasoids and the Mongoloids almost certainly became differentiated from one another somewhere in Asia, and the Caucasoids subsequently spread to the whole western part of the continent, and thence into Europe and north Africa.

Unfortunately knowledge of the early history of this process is limited by the paucity of relevant skeletal remains. It is not until the very end of the palaeolithic period, when the two major races were already well differentiated, that we can form anything like a complete picture of the movements of populations concerned, and even this is mainly limited to the caucasoid peoples. We do not know whether the neolithic revolution in eastern Asia was derived from the west, or was wholly or in part an independent development. Indeed, we cannot exclude the possibility that some of the significant developments which we regard as original to the western (and caucasoid) development of the neolithic revolution, were derived in some way from the far east, where they appear almost as soon as, if not sooner than, in the west.

Around 10 000 BC rapid cultural developments were taking place among the caucasoid palaeolithic peoples of the Near East, leading first to the development of an advanced mesolithic culture which, as we have already seen, was carried by Caucasoids into northern and eastern Africa.

This was rapidly followed by the neolithic revolution, of which the Near East was one of the major world centres, if not the sole one.

The subsequent development of neolithic cultures and their successors in the the Near East itself, and the spread of the neolithic revolution across Europe are fairly well documented, but we know far less about the spread of the latter eastwards into the rest of Asia. Here is a field where archaeology certainly has much to reveal.

In Jericho, at least, we can watch the various techniques characteristic of the neolithic revolution developing and it was certainly from the Near East and the neighbouring lands that, directly or indirectly, much of western Asia, as well as Europe, received its neolithic culture.

In India there are indications that the neolithic culture entered from the north-west (i.e. the 'Near East' of Europeans), and that it was carried by a caucasoid people who came into contact with the late palaeolithic hunting and gathering australoid peoples who were the ancestors of many of the present tribal populations. Here there is archaeological evidence of palaeolithic cave-dwellers living alongside peoples who had reached the neolithic stage of culture.

Neolithic settlements dating from the fourth millennium BC are found in Baluchistan, and the neolithic revolution clearly spread from thence to the Punjab.

The first clear picture that we have of civilization in the region comes from the excavation of Harappa, Mohenjo-Daro and other cities of the Punjab. These were contemporary and comparable with the cities of the Nile Valley and Mesopotamia. Technically they reached in about 2000 BC a similar and even in some respects a higher level of culture, but they have so far yielded few written documents which might ultimately serve to construct a history, and the script in which these few were written has not yet been deciphered. We do not even know to what main family the language belonged.

There is however little doubt that the languages of the neolithic revolution in the region as a whole belonged to the Dravidian family and came from the north-west. We have one living piece of evidence of this in the persistence to the present time of a Dravidian language spoken by the Brahui of Baluchistan. These languages are otherwise confined to the more southerly parts of India. It seems probable that they spread not only with the neolithic culture but with the caucasoid human type and perhaps more specifically with something near that special type which we now associate with the Mediterranean area.

Meanwhile, in south-western Asia, in caucasoid peoples of a somewhat more robust type, new groups of languages (and cultures) were developing. The one which immediately concerns us, and which ultimately became the most widespread of all, was the Aryan or Indo-European group.

The Harappan civilization came to an abrupt end about 1750 BC, clearly as the result of foreign invasion, dramatically attested by the unburied dead found by archaeologists in the streets of Mohenjo-Daro.

This probably marks the beginning of the next cultural phase, initially in the Indus valley, that of the Aryan invaders. They are best known from their great historic epic, the *Rigveda*, transmitted orally for over 2500 years, and not reduced to writing until the fourteenth century AD. In it we find the story of the invasion and overcoming of the previous inhabitants, the 'Dasas'. There are some difficulties in the reconciliation of dates, but it is on the whole probable that it was these Aryans who conquered the Harappans and brought their high civilization to an end. The language of the invaders was thus Indo-European and we have no evidence of the existence of any languages of this family before their arrival.

Thenceforth the history of the Indian region is essentially the story of the Aryans as they penetrated and introduced their language and customs to the greater part of the region. There have, of course, been many subsequent invasions but the numbers of invaders who were added to the general population appear to have been relatively small.

We have seen that during the historical period, from the Harappan and Aryan penetrations to the present time, carriers of a particular culture or of a particular language have usually been of a common genetic stock and have shown certain physical characters in common. We must not however assume that this was always the case, and a study of blood groups is one way of testing a hypothesis which has perhaps been too easily accepted in the past.

We thus see that the peoples of the Indian region fall broadly into three zones—tribal peoples, of australoid physical type, living in pockets chiefly in the south, the Caucasoids of slender type and with rather dark skins, mostly speaking Dravidian languages, occupying the main southern part of the region, and more robust Caucasoids, with paler skins and speaking Indo-European languages, in the north.

We have already seen that many of the tribal peoples have their own characteristic blood-group picture, very different from that of the Caucasoids. Caucasoids everywhere have certain blood-group characteristics, which are of course shared by those of India, notably the presence of the A_2 and d genes. A rather high frequency of MS is shared by most but not all caucasoid peoples.

The Dravidian-speaking peoples of the south of India do possess the caucasoid features just mentioned, but the data for most genetic systems are inadequate for a detailed comparison with the more fully documented peoples of the north. It is hoped that this lack will ultimately be remedied, since these peoples may prove on archaeological and historical grounds to be one of the earliest differentiated branches of the caucasoid 'race'. Their general resemblance to the peoples of south-west Asia and the Mediterranean region is apt to make us look at them as an outlier of this better known group of peoples.

The A_2 gene is present, but at a lower level than in the north. The total frequency of the A gene is below 25 per cent, and that of the B gene about 20 per cent, but it falls gradually southward to the tip of the peninsula and on into Sri Lanka. The frequency of the M gene is highest in the extreme south, falling to the centre but rising again somewhat in the Aryan-speaking north. There are few data on the S gene frequencies in non-tribal populations in the south. We possess, also, very few data on the remaining blood-group systems and other polymorphisms of the non-tribal (i.e. caucasoid) peoples of the south of India. Fuller data on these might throw some light on the ancestral relations between the Dravidian and the Indo-European speaking peoples of the sub-continent.

As we go further north, into and across the Indo-European-speaking zone, we find only a little variation in *A* frequency, but a steady rise in *B* and a fall in *O*. The typically caucasoid gene A_2 is of course present.

The very high frequencies of *B* on the northern confines are shared by the mongoloid peoples to the north and east, suggesting the possibility that a high frequency of *B* is a selective response to some shared feature of the enviroment, such as conditions favouring a particular type of infective disease.

The *M* gene has a rather high frequency throughout the Indian region. In the north its highest frequencies occur in a band just short of the northern mountain boundaries. The frequency of the *S* gene is mostly around 30 per cent, and as in nearly all caucasoid populations, *MS* is greatly in excess of *NS*.

In the Rh system it is difficult to see any definite trend in the frequency of *d* throughout most of the region, but there is a sudden fall as we go north-eastward into regions inhabited by peoples who are partly or wholly mongoloid, in Nepal, Bhutan, and Assam. There is a general northward fall in the frequency of *CDe* and a corresponding rise (common throughout Europe and Asia) in the frequency of *cDE*. There is, however a particularly high frequency of *CDe* in the lower Ganges basin.

South-west Asia

We have already seen, in broad outline, how the neolithic revolution spread from Palestine to India, and that this was carried by caucasoid peoples. It is, however, by no means clear how far these past carriers are related to the present inhabitants of the intervening area. Moreover, though we do not know what language these neolithic pioneers spoke, it may have been a Semitic one, and it certainly was not an Indo-European one, for both in the Indian region and in the Near East itself, the arrival of the Indo-European speakers is a clear-cut event taking place thousands of years later.

Ideally, if we had sufficient data for all the countries concerned, and if we knew how the present inhabitants were related to the early neolithic, bronze-age, and iron-age peoples, we should like to trace the blood-group characters, first of the early neolithic peoples as they spread eastward into Asia, and then those of the Indo-European and Turkish-speaking peoples as they moved southward and westward into the region we are now considering.

Because, however, there have been such great population movements since the neolithic period, and indeed since the beginning of the Christian era, in Palestine and the surrounding countries where the neolithic revolution began, it is difficult to relate recorded history to the genetic anthropology of the inhabitants. In particular the term 'Arab' must be used with great caution.

The best defined 'Arab' population is that of Saudi Arabia and the rest of the Arabian peninsula and, though its origins may go back at least in part to Palestine, and though it is likely to be largely a product of evolution in the isolated and largely desert peninsular area, its characteristics are so relatively uniform, so well studied, and so sharply defined, as to make it a convenient reference population for comparison with the neighbouring peoples.

The peninsular Arabs are remarkable in their high frequencies of the genes *O*, *M*, *S*, and *K*. The lines of equal frequency (isogenes) for these four genes form closed curves around Arabia and they are clearly long-established characteristics of these Arabs, though they are but exaggerations of essentially caucasoid features. The possible special significance of high *O* frequency in isolates is discussed on p. 80. However the Arabs also show frequencies higher than anywhere in Europe or the rest of Asia of the whole range, already discussed, of the African or negroid marker genes. I think however that we are justified in looking upon these as secondary features, due to comparatively recent hybridization, mainly with slave concubines. One indication of the independence of the two sets of characteristics is found in the more isolated population of the island of Socotra which, though very near to Africa, shows high frequencies of *O*, *M*, and *S* similar to those on the Arabian mainland, but much lower levels of the African genes.

As we shall see when we come to look at the Jews as a whole (p. 95), the Yemenite Jews differ from all other Jewish populations in the resemblance of their blood group frequencies to those of the Yemenite Arabs, and those of the peninsular Arabs generally. In the early years of the Christian era, before the advent of Islam, southern Arabia was for long a kingdom under Jewish rule, which ended with the coming of the followers of Mohammed. Many Jews were forcibly converted, and to a large extent the present southern Arabs, though genetically indigenous, may be regarded as the descendants of converted Jews.

A special group of Arabs are the Bedouin of the Sinai peninsula. These are highly inbred populations whose blood-group frequencies are even more extreme than those of the main population of the Arabian peninsula. The Sinai Bedouin fall into two distinct groups, the Towara of somewhat mixed origins and the Jebeliyah who claim descent from serfs from the Balkan area, presented in about 500 AD by the Emperor Justinian to the monks of the monastery of St. Catherine on Mount Sinai. Though now Muslims, their descendants continue to serve the monastery, and are hence excluded from intermarriage with the Towara.

The Towara resemble other Arabs in their high frequency of *O*, but they have an even higher frequency of *K*. The Jebeliyah have less *O* (and much more *B*) than the Towara but more *M* together with 18 per cent of *K*, the highest frequency known. They also have very high frequencies of several of the African marker genes. Both populations have essentially Arab blood-group pictures, clearly considerably modified by genetic drift. The Jebeliyah show little if any evidence of their European origins, but one may suppose that, because of their alien origin and their association with a Christian monastery, their exclusion from intermarriage with the local Bedouin led to considerable interbreeding with African slaves.

As we move north from the Arabian desert, we meet populations in which we might hope to find genetic features related to the origin of the neolithic revolution. We do indeed meet such features elsewhere (pp. 73–4) but it must be remembered that these lands are more fertile than those surrounding them, and hence more desirable. They also permit easier movement, and they lie near the junction of three continents. These lands have thus for many thousands of years seen great migrations of peoples and the passage of many armies. This is dramatically illustrated at the narrow pass between the mountains and the sea, at the mouth of the Dog River in Lebanon, where at one glance the eye can see the numerous commemorative tablets set up by a succession of victorious commanders who for the time being held the pass, from the time of the Pharaohs to that of the Second World War.

We would dearly love to know the blood groups frequencies of the people of the Near East who originated the neolithic revolution, or of their descendants, but unfortunately this is an area where data are both scanty and confusing. However, perhaps the most likely candidates are the settled Arabs of Jordan, the Palestinian Arabs, and the Muslims (rather than the Christians) of Lebanon.

There is a broad general resemblance between these peoples, with considerably lower frequencies of *O* and of *M* than are found in the peninsular Arabs (and in the Jordanian Bedouin). Further east, in Syria and Iraq, data remain scanty and the chief change detectable is a slow eastward rise of *B* frequencies towards the very high values found in the Indian region. The peoples whom we have just discussed mostly speak Arabic, a Semitic language. Further north and west languages belong mostly to the Indo-European and Turkish groups, and, in so far as languages are an indication of physical descent, we might expect to find genetic evidence for a relationship between the Indo-European speakers of Iran and adjacent areas, and the speakers of languages of the same group in India. There is indeed only a slow genetic change as we move from Iran into the Indian region, chiefly a rise in *B* and *M* frequencies, but there is no marked contrast between the Indo-European speakers and their neighbours, so perhaps south-west Asia has been one vast melting pot from which radiated not only genes but languages and culture.

Indo-European languages are today spoken throughout most of Europe, as well as of northern India, and in large parts of western Asia (quite apart from areas of comparatively recent European colonization and settlement). They appear to have originated, probably during neolithic times, in south-central Russia. Between 2500 and 2000 BC they spread southward into South-western Asia where we find them especially in Iran and Afghanistan. It is certain that this spread represents a substantial population movement but we do not know to what extent the immigrants or invaders intermarried with the previous inhabitants of these countries, or of the Indian region. However, as we have seen, the present speakers of these languages are on the whole of a more robust physique than the Semitic speakers who live to the south of them.

The Turkish family of languages are considered to belong to the same super-family as the Finno-Ugric family which we shall meet in Europe. The Turkish languages appear to have originated in Central Asia but are now spoken mainly in Turkestan, just east of the Caspian Sea, and in Anatolia (Asia Minor). Turkish speakers have considerably higher frequencies of the *A* gene than their neighbours in Asia, but the situation is complicated by the Armenians, Indo-European speakers who entered Anatolia from the Balkan Peninsula at about the beginning of the Christian era. They at one time constituted a substantial proportion of the population of Anatolia (modern Turkey) but, as Christians, suffered persecution and massacres so that large numbers of them fled and became dispersed throughout the world. Modern Armenia, in the Soviet Union just west of Anatolia, is inhabited mainly by ethnic Armenians. Both here and in communities elsewhere the Armenians have exceptionally high *A* gene frequencies, around 35 per cent. The Turkmen of Turkestan have high frequencies of *A* but perhaps the even higher frequencies of the Turks of Anatolia are contributed in part by Armenian admixture. The Turkish-speaking Uzbeks of Afghanistan have a high *A* frequency but this is not shown by the data recorded for those of Uzbekistan in the Soviet Union.

As we proceed eastward, both north and south of the Soviet Border, we find (apart from some Turkish populations) a fairly steady level of the *A* gene, between 20 and 25 per cent, but a steady rise of *B*.

South of the Soviet Union, among the Kurds of Iraq and Iran, in the Persians and other peoples of Iran, and in Afghanistan and Pakistan, this eastward trend towards high *B* frequencies is shown by populations which are almost entirely caucasoid, but in the Soviet Union it accompanies an increasing mongoloid component in the population.

Data for the region which we are now discussing are very scanty for systems other than the ABO. There is a slight eastward rise of frequencies of the *M* gene and the *CDe* and *cDE* haplotypes of the Rh system, and a fall in *cde*.

It will be seen that in south-west Asia as a whole, so important in the cultural history of mankind, blood groups have not hitherto contributed greatly to the unravelling of the relationships between the various populations. This is certainly due mainly to lack of knowledge, but probably also to a genuine low level of genetic diversity.

Further north the western part of Soviet Asia is largely inhabited by populations which have fairly recently emigrated from Soviet Europe, but in the extreme north we find the Samoyed reindeer-herding peoples who speak languages of the Finno-Ugric family and probably migrated from the west thousands of years ago. Physically they become progressively more mongoloid as one follows them eastward. They mostly, like the European Lapps (who also herd reindeer and speak a Finno-Ugric language), have a very high frequency of the *N* gene. It is tempting to consider the possibility of their forming a connection with the Ainu with their high *N*, but if must be remembered that even the most easterly of the Samoyeds are still separated by some thousands of kilometres from the Ainu.

The story of the caucasoid peoples of Asia reads like an unfinished chapter, which it indeed is. Many of the themes which we have begun to discuss will be taken up again as we trace the migrations of Asian peoples into Europe.

5. Europe

If man originated in Africa, there is little doubt that it was from Asia that he entered Europe, which was thus the last of the three Old World continents to be populated. The facts of archaeology support this conclusion, though new discoveries are gradually pushing back the age of the first Europeans. Studies of such early events have a great importance of their own, but they have relatively little bearing on the composition and origin of the modern European peoples, most of whose ancestors have almost certainly entered in neolithic and later times. It must however be realized that by the end of mesolithic times the climate of Europe was comparable with the present one, and most of the continent was inhabited as thickly as mesolithic techniques would allow, by men who were almost certainly all of caucasoid type.

In the previous chapter we left the peoples of south-west Asia, with their neolithic culture, ready to enter Europe. The progress of the neolithic revolution across Europe can readily be traced by archaeological means, and dated by radiocarbon estimations. One question which immediately arises is to what extent the revolution was spread by population movement, and to what extent by culture contact, associated with interbreeding between the newcomers and the previous inhabitants. Various writers, especially Bodmer and Cavalli-Sforza (1976), have stressed the fact that neolithic agriculture enabled a given area of land to support a population many times as great as would the hunting and gathering methods of their mesolithic predecessors. They have concluded that the revolution was a self-propagating one spread by a rapidly expanding population, the pioneers in any newly cultivated area being supplied by the rapid multiplication of their predecessors to the east. Yet interbreeding there must have been, and one question which arises is whether this has obliterated the distinction between the two populations, or whether, and how far, we can still distinguish between the descendants of the neolithic invaders and those of the older mesolithic population.

Several lines of approach, linguistic, physical, and genetic, converge to suggest that we can indeed do so. The linguistic argument is at first sight the simplest, but we must as always recognize that languages are not inherited but culturally transmitted and are thus not necessarily indicators of physical ancestry. Most of the inhabitants of modern Europe speak languages of the Indo-European family. Languages of the Finno-Ugric group, and Turkish, the latter still spoken by some Balkan populations, certainly entered central and western Europe long after those of the Indo-European family. Whatever may have been the languages spoken by the original neolithic peoples of Europe, most of their descendants now speak Indo-European tongues. There are, however, a few small pockets of people, physically, and as we shall see genetically, distinct, who speak languages which were certainly there before those of the Indo-European family. These are Basque, spoken in northern Spain and south-western France, and a large number of languages spoken by small relict populations in the valleys of the Caucasus Mountains. Some authorities regard the latter languages as related to Basque, but in view of the many thousands of years which must have elapsed since there was any possible direct communication between the two sets of populations, the connection cannot be a close one. A line drawn between the Caucasus and the Basque country except where in crosses the Black Sea, passes mainly through mountainous territory, to the south of which the populations of Europe are largely of the slender Mediterranean type, similar to the peoples of the 'Near East' whence the neolithic revolution originated. The peoples to the north of the line are on the whole of a more robust type resembling those of Iran and the neighbouring lands where the Indo-European languages originated. This suggests that the neolithic cultures were brought into Europe by two population streams, one corresponding to the present Mediterranean peoples and the other to the peoples of central and perhaps northern Europe, leaving pockets of the pre-neolithic population more or less intact between the two streams.

The blood groups of the Basques

Before we begin an examination of the very varied ethnic, physical, and genetic composition of the peoples of Europe as a whole, we shall examine the more limited implications of a comparison of the Basques with the rest. A study of Basque blood groups was in fact the origin of the hypothesis just discussed.

To understand the background of this study we must look into one of the biological consequences of the genetics and immunology of the Rh blood groups. When it became known that the cause of haemolytic disease of the newborn was an incompatibility between the Rh groups of a mother and her child, Haldane and Wiener independently pointed out the implications of this for the population genetics of the Rh groups. A mother whose child or children suffer from haemolytic disease is almost invariably of the Rh genotype *dd* and lacks the *D* gene and the corresponding D antigen. Any D-positive children she may have must be of genotype *Dd*, having inherited the *d* gene from her and the *D* gene from the father, and it is such children that may suffer from the disease. Thus deaths from this formerly highly fatal disease must in the aggregate cause the destruction of equal numbers of *D* and *d* genes in the population as a whole. Therefore, whichever gene was at the outset the rarer of the two should, in the long run, disappear completely from the population.

Most of the peoples of Europe include about 16 per cent of 'Rh-negatives' (i.e. persons of genotype *dd*); simple calculation shows that this corresponds to a *d* gene frequency of 40 per cent and a *D* of 60 per cent. Thus the *d* gene must, however slowly, be on its way out. One possible explanation for the existence at present of such an unstable situation is that the main population of Europe is the result of the mixing of two populations not many thousand years ago, one with a very high frequency of *D* genes and another with a very high frequency, necessarily above 50 per cent, of *d* genes, and that selection is still in progress. In trying to interpret the hypothesis in terms of known populations it was easy to suggest candidates, in western Asia and elsewhere, for the high *D* population, but no population was yet known with more than 50 per cent of *d*. It was then shown that the Basques, not previously Rh tested, did have about 60 per cent of *d* genes, and they still remain the only large population known with more than 50 per cent of *d*. They had other characteristics which fitted them for the role of descendants of the main pre-neolithic population of western Europe, if not of Europe as a whole. They have the lowest *B* frequency in Europe (and perhaps had originally no *B* at all), they have some skeletal resemblance to the late palaeolithic inhabitants of western Europe and, as already mentioned, they are the only west European population speaking a non-Indo-European language. In addition we find, over a very large part of south-western France, a rather high frequency of *d* genes and a low one of *B*, and a considerable number of Basque place names, all of which may be regarded as indicators of a former wider distribution of the Basque population. As we then go eastward, up to the eastern confines of the continent, there is a steady fall in *d* and a rise in *B*. It is therefore suggested that as the bearers of the neolithic revolution progressed across Europe they came up against people akin to the modern Basques. The latter were, however, not exterminated but in the course of time hybridized with the newcomers, giving rise to the present peoples of Europe. After an interval of six thousand years or more it is possible to reconstruct only a rough approximation to what actually happened. In particular, both the newcomers and the people whom they met were certainly more complex than is suggested, and it may be that we shall yet find evidence of some of these complexities.

The nations of continental Europe

We have up to now treated the neolithic newcomers from Asia into Europe as at most two major populations, and seen how we can account for some features of the blood groups of their descendants by assuming the interbreeding of a majority of invaders with a minority population which was there already. Since that series of events, however, some six thousand or more years ago, there has been ample time for large and complex population movements both within Europe and over the border from Asia, and there is ample historical evidence at least for the more recent of such movements. The results of these can be seen in the present complex mosaic of physical types, and of languages, which sometimes do and sometimes do not correspond with one another. We must now try to see how far blood-group distribution can help to untangle the pattern.

In nearly all parts of the world the ABO blood groups show a much more complex mosaic than do those of the other systems, and, because of the very detailed blood group surveys which have been carried out, this is seen more clearly in Europe than in the other continents. The pattern is to be regarded as the result of natural selection related to the harmful effects of particular climatic and other local features of the environment. We still do not know at all precisely which environmental features are involved, but these almost certainly express themselves by tending to cause particular diseases, to each of which people of one ABO group are more susceptible than those of another. Some of the possible mechanisms involved are discussed on pp. 000 and 000, and I have treated the subject at length in a separate book (Mourant *et al.* 1978). The times required for substantial frequency changes in the ABO groups appear to be of the order of one to two thousand years. Over periods of only a few hundred years there can be no doubt that populations retain their blood-group characteristics, and this is demonstrably true even when they have moved to a new environment. Since many of the great population movements in Europe took place about a thousand years ago, their effects on ABO distribution ought still to be detectable, whereas events of six thousand years ago which, as we have seen, seem still to be detectable in terms of Rh distribution, are probably completely blurred on the ABO map. The MN groups, for which we have ample data for the whole of Europe, are probably selected even more slowly than the Rh groups.

The Rh groups, as we have seen, show only a gradual change in frequencies as we follow them across Europe. There is a steady rise in *d* frequency from about 30 per cent in the east to 45 per cent in the west, and to 50 per cent or more in the Basques and a few other small and peripheral groups. At any given longitude, *d* frequencies are lower in the Mediterranean area than further north. Cutting across the generally east–west gradient of *d* and its allele *D*, the *CDe* haplotype is highest in the south and *cDE* in the north, a trend which is visible throughout Asia as well as Europe, so that this may perhaps be the result of natural selection related to climate. Frequencies of the MN blood groups are very nearly uniform within Europe. Most western populations have *M* gene frequencies between 50 and 55 per cent, and there is a slow and fairly uniform eastward rise to just above 60 per cent, a trend which continues into Asia where considerably higher frequencies are found. The distribution of the *S* and *s* genes (which are closely linked to *M* and *N*) is not sufficiently well documented to enable them to contribute substantially to the classification but the fact that in Europe *S* is mainly linked with *M* and *s* largely with *N* may be additional evidence to that discussed on p. 130 for the European population having undergone important hybridization in the last few thousand years.

For most of the other blood-group systems gene frequencies are nearly uniform within Europe, so that they cannot be used as a means of classifying populations. However, for most systems, including ABO, Rh, and MN, European averages differ considerably from those of the population of the other continents. The distribution is not simply one between Caucasoids and non-Caucasoids, for the Caucasoids of south-west Asia, and even more so those of India, differ widely from those of Europe. This is likely to be due partly to original genetic differences between the neolithic and earlier populations entering Europe, and those left behind in Asia. The migrant populations must also have been modified by natural selection as a result of entering the new environment. The differences are unlikely to be due to any great extent to genetic drift, because of the large sizes of the populations involved.

The ABO groups

In spite of the fine grained chequer-board pattern that dominates the ABO maps of individual countries, world maps show an overriding general pattern. Firstly, most European populations have more than 25 per cent of *A* genes, and this trend is continued into the Turkish-speaking countries of south-west Asia. Outside this area there are few populations anywhere in the world with such high *A* frequencies. Within Europe there is a western zone of high *O*, and relatively low *A* and *B* frequencies. This zone is discussed later, with particular reference to processes of natural selection and genetic drift. Most of continental Europe falls into two main zones, a western one of high *A* and an eastern one of relatively high *B*, rising steadily eastward. The high *A* peoples mostly speak Indo-European languages other than Slavonic. Peoples with over 10 per cent of *B* genes nearly all speak Slavonic and Finno-Ugric languages. The main Finno-Ugric speakers, Finns and Magyars, have somewhat higher *A* frequencies than the Slavonic speakers. According to some authorities the Turkish languages are related to the Finno-Ugric ones and, as we have seen,

Turkish speakers outside Europe share this combination of high *A* and *B*.

In the extreme north of Europe, and continuing eastward into Asia is a particularly interesting group of Finno-Ugric speakers, the Lapps and the Samoyeds, who may or may not be related to one another.

The Lapps are an ethnic group living in northern Norway, Sweden, Finland, and north-western Russia. Some are reindeer herders and some are fishermen. They are on the whole of caucasoid appearance but with a very slight mongoloid tendency. They speak a Finno-Ugric language closely related to Finnish, but regarding this there are two points of view. One is that they formerly spoke a non-Finno-Ugric language now lost, and adopted Finnish which in the course of time became modified. The other is that their original language was a Finno-Ugric one, but belonging to the Ugric sub-family; this however became modified by the very large-scale adoption of words from the language of their neighbours, the Finns. This distinction is an important one with regard to their possible relationship to the Samoyeds.

The Lapps have been very thoroughly studied from the point of view of blood groups. They are almost unique in their high frequency of *A* and totally so in having the highest A_2 gene frequency known, reaching 42 per cent in one group tested. In this respect they are super-Caucasoids, for the A_2 gene is almost entirely confined to caucasoid and negroid populations in whom it is however mostly below 5 per cent and only extremely rarely exceeds 10 per cent.

Most Finnish Lapp populations have over 50 per cent of *N* genes; the Norwegian and Swedish Lapps have a higher frequency of the *NS* haplotype than any known populations except the Ainu. In the Rh system they have one of the lowest *d* frequencies in Europe, and they have about 12 per cent of the otherwise rare C^wDe haplotype, which they appear to have passed on to their Scandinavian neighbours. The Lapps thus differ widely from all other known populations, with a tendency towards Asiatic values for some characters (high Fy^a and PTC taster gene) but not for others (ABO, MNSs). Their origins and relationships are thus something of a mystery, and it has even been suggested that they are descended from a population which survived the last ice age isolated in an unglaciated area (which certainly existed) in the extreme north of Norway, beyond the great European icesheet.

It has, however, recently become possible to make limited blood group comparisons with another group of Finno-Ugric speakers, the Samoyeds, reindeer herders who migrated eastward into the north-western parts of Asiatic Russia. These have rather low frequencies of *A* and high of *B*, but they do resemble the Lapps in having exceptionally high frequencies of *N* and they could thus conceivably provide a connecting link between the otherwise isolated Lapps and Ainu, both also with very high *N*.

The high-O regions

Some of the earliest regional studies of ABO distribution were carried out in the British Isles, and showed a general increase of *O* frequency from relatively low values in southern England, to higher ones in northern England, Wales, Scotland, and Ireland. This suggested that the Anglo-Saxons had relatively high *A* levels, and that *O* increased as the proportion of 'Celtic' ancestry increased. It was known, also, that *A* frequencies were even higher, and *O* lower, in Scandinavia than even in southern England. It was then found that the Icelanders had high *O* frequencies, close to those found in Scotland and Ireland. The historical Icelandic sagas seemed to show that most of the original inhabitants of Iceland came from Norway, and a minority only from the British Isles. The blood-group findings appeared to reverse this. Then, over a period of years not only were further confirmatory ABO tests done, but tests were done for further blood-group and other genetic systems. Some appeared to confirm and others to contradict the conclusions drawn from the ABO tests. Finally, tests were done on some 2000 Icelanders, mostly of precisely known birth-places within Iceland, for some twenty systems. The results of the tests were then compared with the results of similar tests on the populations of the separate countries of the British Isles and Scandinavia, and of several other European countries. A large quantity of data was fed into a computer, using a highly sophisticated programme, and it was anticipated that the result would be a clear-cut indication of either a Scandinavian or British origin, or perhaps a precise estimate of the proportion of genes derived from each of the two sources. Neither of these was found to be the case. The Icelanders showed a very marked difference from the populations of all other European countries, British, Scandinavian, and other, and even wide differences between the regions within Iceland itself.

Since there is no doubt that the original colonists of Iceland came almost exclusively from Scandinavia and the British Isles, there must have been great changes in the island gene frequencies since the colonization. Natural selection may have played a part, but there can be little doubt that we are witnessing what are mainly the effects of genetic drift due to severe epidemics, volcanic eruptions, and volcanically initiated floods. These have at various times over the centuries reduced the populations of different regions, and of Iceland as a whole, to levels where great accidental fluctuations of gene frequencies were possible, and such fluctuations seem indeed to have occurred so that, as we have seen, the frequencies observed at present bear little relationship to those of the original colonists.

The ABO system is, however, perhaps a special case, and the resemblance of the Icelanders to the Scots and the Irish, in all these peoples having high frequencies of the *O* gene, may not be due simply to the coincidence of three separate random fluctuations. If we examine the frequencies of a number of blood group genes, not only in the Icelanders, but in as many as possible of the isolates scattered in and near Europe on islands, on mountains and in deserts, we find in nearly every case that several genes show frequencies outside the normal range found in larger populations. However, for most systems the departure from the normal is in different directions in different populations, high *D* or high *d*, high *M* or high *N*, as would be expected if these resulted from genetic drift. But, for the ABO system the departure is almost invariably in one direction, that of high *O*. Almost the sole exceptions are the Lapps (if they are few enough to be considered as an isolate at present or in the past) and the Flittas of Oran in Algeria, both with very high A_2. A raised *O* frequency is so almost invariably found in isolates that it is unlikely to be accidental. There are two possible explanations. One is that most of the isolates represent pockets left over from some early population, which itself had a high *O* frequency. This could explain some of the observations, for a number of the western isolates or near isolates are populations which speak or spoke Celtic languages. The phenomenon is however much too widespread for this to be the main explanation and we must rather look to natural selection.

Such a mechanism is in fact known to exist, and will be discussed in greater detail in Chapter 9. There is a tendency in all populations for a higher proportion of A and B than of O fetuses to die of haemolytic disease of the newborn or to be aborted spontaneously, so that all populations should gradually, in the course of ages, show an increasing proportion of surviving O children, and hence of O adults. It is suggested that in populations isolated by the sea and mountains this tendency operates almost unopposed, but that in populations having free communication with surrounding areas certain infections tend to spread which affect O persons on the average more severely than A or B ones, so that a selective equilibrium is reached at higher *A* or *B* frequency levels than in the isolated populations. Provided only that the *O* gene is present (and we know of no population without *O*) this pair of opposing selective forces should continue to operate in spite of the lapse of time, and of the effect of genetic drift on other blood group systems. Drift will indeed affect the ABO system too, but any extreme frequencies which may-develop through its chance operation will tend to be brought back to equilibrium levels within a few generations by the constant pressure of natural selection.

In Table 5 we can see the almost constant appearance of high *O* frequencies in the isolates of Europe, the Near East, and North Africa, and the erratic fluctuation of the genes of the other widely studied systems,

TABLE 5
Blood group gene frequencies in isolates in and near Europe

Isolates	Gene percentages					
	A	*B*	*O*	*M*	*d*	*K*
Icelanders	20	5	75	58	37	5
Irish (Republic)	17	7	76	57	43	4
Lapps	37	9	54	52	16	1
Basques	24	2	74	54	56	5
Béarnais	24	4	72	49	59	3
Corsicans	22	3	75	65	35	4
Sardinians	20	7	73	75	22	3
Walsers	21	5	73	51	41	8
Bergamasques	24	6	70	56	43	5
Valle Ladine	20	3	77	78	56	
Svani (Caucasia)	23	7	70	65	41	
Saudi Arabians	14	11	75	72	25	6
Towara Bedouin	16	9	74	52	31	13
Jebeliya Bedouin	12	26	62	66	54	18
Ait Haddidu Berbers	7	5	89	24	23	4

MN, Rh, and Kell. The more extreme of these fluctuations are found, however, in small isolates. The larger semi-isolates, such as the Scots and the Irish, show them in lesser degree, and there are clines connecting the frequencies found in these isolates, for all systems including ABO, with the more central and less isolated populations with which they are in contact, such as the English and the French.

In intermediate zones, such as northern England, Brittany, and Normandy, we find intermediate frequencies of the ABO genes. The populations of these zones probably show the result of mixing over the centuries between the isolates, with their frequencies developed in the manner already discussed, and the more central populations whose development we have also discussed. As far as the ABO system is concerned (and if we are right regarding the effects of natural selection) the selection pressures themselves in this intermediate zone will also tend to set up equilibria at intermediate levels.

However, in England and France, where numerous persons of more or less precisely known birthplace have been tested for ABO, the detailed analysis of very large numbers of observations (e.g. Kopeć 1970) have shown that, superimposed upon the general cline there are many minor fluctuations which probably represent the selective effects of local variations of the environment, as well as the results of migrations that have taken place sufficiently recently not to have been obliterated by genetic drift and selection; that is to say within the last few hundred years, or perhaps as much as a thousand years.

Cornishmen, Bretons, and Normans

When the Roman armies withdrew from Britain in the fifth century AD, and the English or Anglo-Saxons began to occupy the country, large numbers of Britons fled to north-western France, mainly to what is now Brittany, but also to Normandy, where towns and villages now called Bretteville, and churches dedicated to British saints, are evidence of their presence. Later, the latter area also received Scandinavian settlers, the Northmen or Normans.

The peoples of western Brittany have less than 25 per cent of *A* genes and the area thus falls within our north-western high *O* zone, but Normandy and eastern Brittany also have shown more O and less A than the neighbouring provinces. Since the Norman invaders probably, like modern Scandinavians, had high A frequencies, it is likely that the high O throughout the region is a result of British settlement. Devon and Cornwall were for a long time a stronghold of the British, and Cornwall remained partially Celtic speaking until the eighteenth century, yet ABO frequencies are typically English rather than Celtic. There is little doubt that this is due to the relative openness of the area to traffic from the east, but whether it is English genes or English epidemics that are responsible it is difficult to say.

The high *A* region of Europe

We have already seen that over much of Europe the frequency of the *A* gene is above 25 per cent, and that of *B* below 10 per cent. We can therefore arbitrarily define an extensive high *A* region as lying east of the 25 per cent *A* isogenic line, and west of that marking 10 per cent of *B*. The latter line runs down the Gulf of Bothnia and the Baltic Sea, and then enters Germany, bulging westward to include on the high *B* side the middle and upper course of the River Elbe. There is a slight re-entrant marking lower *B* frequencies near the eastern corner of Czechoslovakia, then a westward bulge takes in the whole of that country on the high *B* side. The line then crosses the Alps and passes down the length of the Adriatic Sea. The line follows on the whole the linguistic border between Germanic and Slavonic speakers, but deviations suggest that populations may in the past have changed their language one way or the other.

The high *A* area which we are now considering includes France, most of England, western and part of eastern Germany, Switzerland, most of Austria, the Low Countries, and Scandinavia (except Finland). Two highly deviant ethnic groups, the Basques and the Lapps, have already been discussed but the remaining peoples of these countries, despite their well-known physical variability, are very uniform in their blood group frequencies. Though the area was defined in terms of ABO frequencies the latter are some of the most variable features, especially those of the *A* and *O* genes, for the frequency of B is, by definition, relatively low.

The Rh blood groups show little variation throughout the region. On the whole, Rh-negative frequencies are lower in Scandinavia and in the east than in the west, while cDE and the rare $C^{w}De$ tend to increase northward and eastward. There is almost complete uniformity of distribution of the MNSs groups, probably of P (where technical errors may be responsible for some apparent variations), and almost certainly of the Lutheran, Kell, and Duffy groups. The rather scanty data on the inherited plasma groups and red-cell enzyme variants also show little variation. There have been considerable historically recorded population movements in this area in the last two thousand years and we must suppose that, despite variations of physical appearance, there have been sufficient mixings of population to bring about the observed relative genetic uniformity. There are nevertheless a number of known genetic isolates within the area, and others may exist. Well planned local blood group surveys, for which the region is particularly well suited, would certainly reveal features of anthropological interest.

The eastern high B zone

The high-B zone, extending from the 10 per cent *B* isogenic line to the eastern confines of Europe, is inhabited mainly by speakers of Slavonic languages, and languages of the Finno-Ugric family (Finns, Hungarians, Lapps, and some populations in Russia). The Slavonic speakers of central Europe, in Poland, Czechoslovakia, and Yugoslavia differ from their immediate western neighbours mainly in their rather higher frequencies of *B* and of *M*. Their Rh frequencies are typically European with about 42 per cent of the *d* gene. Large numbers of ABO surveys have been carried out on the Slavonic speaking peoples of the European part of the Soviet Union, but the area and the total population are so great that points of known frequencies are very thinly scattered on the map, and it is difficult to distinguish a pattern. There is, however, a suggestion of a series of tongues of high B and low O frequency, based on the northern end of the Caspian Sea, extending westward into eastern Germany and the Baltic countries. One of these seems to point towards the low-O area of Finland, and may mark the north-westward passage of people ancestral to the Finns. Data for MN and Rh are very scanty but there seems to be a gradual eastward rise of *M* and fall of Rh-negative frequencies.

The East Baltic peoples

The blood groups of the Finns have been studied very extensively, especially for the ABO and MN systems, with respect to which they differ considerably from most of their neighbours. Our very slight knowledge of the blood groups of the Lithuanians, Estonians, and Latvians suggests that they are related to the Finns.

The Finns share with the Scandinavian peoples a very high frequency of the *A* gene, but they differ from the Scandinavians in having a much higher frequency of *B* and from the Lapps in having a much higher frequency of *M* and a much lower frequency of A_2. Rather surprisingly the A frequency is lower in western Finland, which is adjacent to Sweden with its high A frequencies, than in the east.

The general M frequency in Finland is higher than is recorded for any other European populations apart from the neighbouring Estonians, and the Sardinians. This is probably related to the rather high *M* frequencies found in Russia and the even higher ones of most of Asia.

A small amount of work has been done on the Komi, the Mari or Cheremiss, and the Udmurts or Votyaks, all Finno-Ugric speaking peoples living in Soviet Russia and supposed to be related to the Finns. As far as their ABO groups are concerned, their *A* frequencies are considerably lower and their *B* frequencies higher than those of the Finns, and more like those of their Russian neighbours. On the other hand the rather high *A* and *M* frequencies found in the Finno-Ugric-speaking Magyars of Hungary, as compared with their immediate neighbours, are probably connected with their relationship to the Finns.

Southern Europe

The lands surrounding the Mediterranean Sea constitute a natural geographical region and the peoples of this area, with a long history of intercourse, both peaceful and warlike, undoubtedly have a considerable degree of kinship with one another. As far as the peoples on the European side are concerned we have already suggested that this kinship arises from their being decended from a single stream of peoples who brought the neolithic revolution from the Near East to this part of Europe, and this would also explain their physical similarity to the peoples of south-west Asia as well as of much of the Indian region. A large proportion of the peoples of this extended region, including also those of the southern shores of the Mediterranean Sea, are of the so-called Mediterranean physical type, rather short in stature, long-headed, with light to medium brown, or even dark brown complexions. In Europe and to a considerable extent also in Asia they speak languages of the Indo-European group.

We shall consider the peoples of the European side of the Mediterranean in order from west to east. There is not here, in the south, a western zone of high O frequencies like that found in the north, but certain populations, especially the islanders of Corsica and Sardinia, have high O frequencies. Otherwise, ABO frequencies are similar to those found in the central or high *A* zone of northern Europe, and *B* gene frequencies above 10 per cent are found almost solely in the Balkan countries and on the Black Sea shores of the USSR.

Variations in frequency of the Rh groups are greater than in northern Europe but there is some regularity in the pattern, which helps considerably in classifying the populations of the region. Frequencies of the Rh-negative type, and of the underlying *d* gene, are on the whole considerably lower than to the north. It has already been suggested that the ancestors of the Basques, with their very high *d* gene frequency of 55 per cent, made a considerable contribution to the composition of the peoples of northern Europe. It will be seen that they have probably made a similar contribution to that of the Mediterranean Europeans. The northern Sardinians have a very low frequency of *d*, only 22 per cent, with a very high frequency, 67 per cent, of the *CDe* haplotype. It is possible to account for the Rh composition of most of the Mediterranean peoples by assuming that their main component has the same Rh composition as the northern Sardinians, and is mixed with north European and Basque components in varying proportions. This would suggest that the Sardinians, at least as far as their Rh groups are concerned, resemble closely the original neolithic invaders of southern Europe. A fourth component, an African one, is needed to account for the raised frequency of *cDe* found in several parts of the area. This haplotype has an average frequency of only 2 per cent in most parts of Europe, but 60 per cent or more in most black African populations. The African component is therefore very easy to recognize in the results of any blood-group survey.

The frequencies of the *M* and *N* genes resemble those found further north, and *M* shows a similar gradual rise from west to east.

The Iberian peninsula

The Basques, who live partly in France and partly in Spain, differ widely in their blood-group composition from all surrounding peoples; they have already been discussed and only incidental mention will be made of them here.

In Spain there is an area of low B frequency including the Basque country but extending beyond it; otherwise ABO frequencies in the Peninsula are relatively uniform and unremarkable, as are MN frequencies. Rh-negative (*d*) frequencies in Portugal are similar to those of northern Europe but in Spain, with two recorded exceptions, they are lower and more typically Mediterranean. One isolate in the Pyrenees, in the Valley of Bielsa, quite separate from the Basque country, shows a similar very high frequency of Rh-negatives, accompanied by other aberrant frequencies typical of a genetic isolate. With a rather high frequency of *O*, already discussed as a specific characteristic of isolates, they have an exceptionally low frequency of *M* and a relatively high one of the *K* gene of the Kell system.

Corsica and Sardinia

The adjacent large islands of Corsica and Sardinia have populations differing very considerably from those of the neighbouring continental areas, and despite different political histories their populations show some marked resemblances, as well as some differences. The frequency of group O is high in both islands, a feature with which we are now familiar in isolates. Unlike the Basques and the people of the Val de Bielsa just described, the Corsicans and Sardinians, but especially the latter, have low frequencies of Rh-negatives and high ones of *M*. It might be supposed that the extreme ABO, MN, and Rh frequencies found in Sardinia and especially at Sassari were the result of genetic drift acting in a small isolated population. However the resemblances between Corsica and Sardinia suggest that the original population developing these traits was not an extremely localized one and we have already suggested (p. 84) that the Rh pattern of the Sardinians is a very ancient one. The ABO frequencies in both islands are indeed similar to those of other isolated populations; we have seen that this does not necessarily imply consanguinity but may be due to processes affecting all isolates whatever their origin.

Certainly, however, history and archaeology show the Sardinians to have kept very much to themselves for thousands of years. Their prehistoric nuraghe or fortified towers, of which hundreds still exist, are well known, and since the time of the Carthaginian occupation the native inhabitants are known to have retired repeatedly to their mountain fastnesses whenever a strong invading force appeared, and recent events show that this tradition still exists.

The Sardinians not only have unusual blood-group frequencies but, like many other Mediterranean populations, they have the genes for glucose 6-phosphate dehydrogenase (G6PD) deficiency and for thalassaemia. As we have seen, these two conditions, which are disadvantageous under most conditions, have long been suspected of conferring resistance to malaria. The latter disease was formerly prevalent in Sardinia. Though it is now rare or absent, it is thought that its genetic effects are still persistent in the island, and this has led to its becoming a very important human genetic laboratory.

In central Sardinia there are a number of adjacent villages, some at low altitude and formerly much subject to malaria, and some at high altitude and with little malaria. A succession of Italian workers, beginning with

Carcassi, showed that the frequencies of all the ordinary blood groups were closely similiar in all these villages, this suggesting that at no very remote period the inhabitants were drawn from the same stock. However the incidence of thalassaemia was found to be significantly and consistently higher in the lowland than in the highland villages. A few years later Siniscalco confirmed this work, but also showed that G6PD deficiency had a similar high incidence in the lowland but not the highland villages. It thus seems that, in the presence of malaria, the carriers of both these genes are selectively favoured, but in the absence of malaria their otherwise harmful effects cause them to tend to die out. It has since been shown directly that G6PD deficiency confers resistance to malaria, but the same has not yet been shown for thalassaemia.

The very thorough genetic background which has been established in Sardinia has prepared the way for other important medical observations, such as a probable association of G6PD deficiency with a tendency to duodenal ulceration.

Italy and Sicily

There is a fairly steady gradient of most blood-group frequencies in Italy from north to south, but it is convenient to consider the country as falling into two main zones, the division lying about 100 km north of the latitude of Rome. Ferrara and the lower Po valley show more southerly characteristics than the rest of the north.

The northern area has relatively high frequencies of the *A* gene, mostly above 25 per cent and rising to about 29 per cent around Milan, Padua, and Piacenza. However, in the area inland from Venice and Trieste, the *A* frequency falls below 25 per cent and in the Alps north of Belluno this characteristic is even more marked in the Valle Ladine isolate, which perhaps formerly occupied a more extensive area. Somewhat similar but less marked features are found in the towns of Bergamo and Brescia, and the area to the north of them.

The frequencies of the *M* gene are somewhat higher in Italy than in most parts of northern Europe, averaging about 57 per cent, yet frequencies are on the whole higher in the north, where they reach 60 per cent locally, than in the south, where Sicily has only 52 per cent.

The frequencies of the Rh blood groups follow a similar pattern to the ABO, with *d* frequencies above 35 per cent in the north and below that value in the south. The two areas of high *O* frequency in the Alpine region also show high frequencies of the *d* gene. Rh frequencies in the country as a whole are, however, typically Mediterranean, with high frequencies especially of the *CDe* haplotype, around 50 per cent.

Thalassaemia and glucose 6-phosphate dehydrogenase deficiency are not uncommon over large parts of Italy. Thalassaemia in particular is most common in the south and in the lower Po valley. A small percentage of Haemoglobin S (sickle-cell haemoglobin) occurs in Sicily where it is presumably of African origin. As we have seen, these three genetic factors are all probably selectively favoured by the existence of malaria and tend to occur in formerly malarious areas.

Yugoslavia

The high B zone of northern and central Europe continues into the south. The peoples of Yugoslavia thus differ sharply from those of northern Italy in having a higher *B* and lower *O* frequency. Despite the ethnic variety of the peoples, ABO frequencies vary very little from one part to another, and average about 28 per cent of *A* and 13 per cent of *B* genes.

Rh blood group frequencies have a north European appearance, with an average frequency of 40 per cent of *d*, but *CDe* and *Cde* tend to be high and cDE low, as in the Mediterranean area generally.

Romania

The population of Romania differs little from that of Yugoslavia in its ABO distribution, apart from a slightly higher *B* frequency. Despite the presence of the Carpathian Mountain barrier, no large ABO frequency variations seem to occur. Rather unusually, the MN groups, as well as the Rh, show greater variation, with *M* gene frequencies rising from west to east from 54 to 62 per cent. For the Rh system, frequencies of the *d* gene vary from 34 to 39 per cent, and possibly to 45, but the last figure could possibly refer to selected blood donors.

Bulgaria

Bulgaria, together with European Turkey and the south-east corner of Yugoslavia, stands out from the rest of south-east Europe in having an *A* gene frequency uniformly a little above 30 per cent. The very few data on MN and Rh show 56 per cent of *M* genes and 38 per cent of *d* genes, figures near those of neighbouring countries.

Greece

Blood group frequencies in Greece differ considerably from those found in the countries with which she shares a land frontier, and are much nearer to those of southern Italy. Not surprisingly, however, in a country crossed by high mountain ranges, and including large numbers of islands, there are marked internal variations. The average *A* and *B* gene frequencies are near 25 and 10 per cent respectively and that of *M* is about 57 per cent. In the Rh system, the frequency of the *d* gene averages about 32 per cent, near to that for southern Italy, but frequencies in the islands tend to be low, down to 24 per cent in the Dodecanese. As usual in the Mediterranean area, *CDe* has a high frequency, and *cDe*, presumably of African origin, averages 6 per cent.

Much work has been done on the distribution of thalassaemia and of glucose 6-phosphate dehydrogenase (G6PD) deficiency, both of which are not uncommon and tend, as elsewhere, to follow the former distribution of malaria. The frequency of G6PD deficiency is especially high, near 32 per cent in males, both in the island of Rhodes and at Khalkidhiki on the mainland. Haemoglobin S occurs sporadically, and there is a concentration of it around Lake Copais in central Greece, in what was formerly a highly malarious area. It might have been expected that this African genetic character would have been accompanied by raised frequencies of other African genes, but this is not the case, suggesting that the Haemoglobin S gene was brought in by small numbers of carriers a long time ago, and then was favoured by natural selection in the malarious environment.

6. The Jews and the Gypsies

Both the Jewish peoples and the Gypsies originated in Asia, and it might seem therefore that they should have been described in the chapter on that continent. However, though, as we shall see, both groups of populations retain abundant genetic evidence of their origins, they have, in the course of their age-long wanderings, acquired genes from Europe and, in the case of the Jews, from Africa as well, and their full genetic picture can be understood only in the light of what we know of the peoples of these other continents.

The Jews

The genetic picture of the Jews is in one sense confused by their early wanderings and ill-recorded hybridizations, and the conversion of non-Jewish peoples to Judaism, but the position is simplified by the undoubted fact that most Jewish populations have now, for some hundreds of years, been closed and highly endogamous communities, so that the population samples which we now study genetically are to a high degree representative of the ancestral populations which emerged into the full light of history in post-Renaissance Europe.

Whatever may have been the realities behind the Biblical story of the wanderings of Abraham, the sojourn in Egypt, and the Exodus, the first clear picture which we have of the Jews is of a settled population in Palestine who probably closely resembled the other populations of the Near East.

It must be remembered that in the time of Solomon the land of Israel extended far to the east of the River Jordan, up to Damascus. After his death the country split into a northern and a southern kingdom, and quarrels and wars between them gave opportunities for conquest to the great eastern powers of Assyria and Babylonia.

In 732 BC Tiglath-Pileser, king of Assyria, captured Damascus and deported many of the inhabitants; in 722 his successor, Shalmaneser V, captured Samaria in the heart of the northern kingdom and the latter's successor, Sargon II, in turn took away many of the inhabitants to 'Halath, Hathor and the cities of the Medes', replacing them by aliens from other subject territories, a strategy only too familiar in modern times. This was the real beginning of the dispersion or Diaspora of the Israelites.

Subsequently, to restore order in Samaria, priests were sent back to reinstate the worship of Jahweh (Jehovah), the god of the land. This was almost certainly the beginning of a syncretic or diluted form of Jewish worship, controversy on the legitimacy of which continues to this day between Samaritans and Jews.

Meanwhile the southern kingdom of Judah continued its rather precarious existence but a generation later the Chaldeans or Neo-Babylonians under Nebuchadnezzar II captured Jerusalem and took many of the inhabitants, especially of the priestly and ruling classes, into captivity in Babylon. Some however, including the prophet Jeremiah, fled to Egypt. This was the beginning of the great Jewish colony in Babylon which, for over 2500 years, until AD 1948, remained one of the largest, and for the greater part of the time one of the most influential, centres of world Jewry. These were the original Jews of the kingdom and tribe of *Judah*, as distinct from the peoples from the northern kingdom who became the 'lost tribes of *Israel*'.

In 538 BC Cyrus, king of the Medes and Persians, having conquered Babylon, allowed exiled Jews to return to Jerusalem, though many preferred to stay in Babylon. In 445 BC Nehemiah repaired the walls of Jerusalem and some years later, probably under Ezra, the temple was rebuilt. During succeeding centuries a succession of external rulers held power over Palestine, which however had a certain degree of autonomy. Even the Romans, who under Pompey took Jerusalem in 63 BC, at first allowed the Herodian dynasty to rule the country, but following a revolt of the Jews in AD 70 Titus, later Roman Emperor, seized Jerusalem from the rebels and, against his orders, the Temple was destroyed. The representation, on the Arch of Titus in Rome, of the triumphal march with the Temple treasures on display, is well known.

This marked the end of any kind of Jewish autonomy, and the beginning of a period of severe repression during which a great many of the Jews who survived left the country, and Jerusalem ceased to be a Jewish city.

Many of the refugees went to Babylon, or to Alexandria in Egypt, but by this time there were Jewish colonies in most countries and large cities in the east Mediterranean area, and in Rome. It was to these that the Jews of Palestine gravitated over the first few centuries of the Christian era, to give rise in due course to the Jewish communities of Europe.

Though a small number of religious Jews remained in Palestine, especially at Tiberias, the religious centre of Judaism moved to Babylon where it remained for several centuries.

The lost tribes of Israel

The Israelites exiled from the northern kingdom ultimately became dispersed throughout south-west Asia, and even as far east as India and China, but most of the communities retained their identity and their religion, and many have continued to speak Aramaic, the Semitic Palestinian colloquial language of the time of Christ. Others speak a Judeo-Persian dialect. Some small communities can indeed plausibly be traced back to the earliest dispersal of all, that of the east bank tribes from near Damascus.

It is only comparatively recently that the existence and origin of these communities have become generally known, and members of many of them have found their way back to Palestine in the great return.

Before that, it was popularly supposed that the 'lost tribes' had indeed disappeared, and so it became possible for romantic historians to identify them with such modern populations as the American Indians and the British, and to apply to these the prophecies if the Old Testament and the book of Revelation.

A numerous and well defined group of Asian Jews is that of the Yemenite Jews from Yemen in Arabia. There is little in their history to set them apart from other Jewish communities of Asia, but their blood group picture is, as we shall see, a rather special one (p. 95). The Jewish community in Egypt, augmented by refugees from Palestine, was also harassed by the Romans, so that many Jews fled westward into other parts of north Africa, where there were probably already some Jewish communities. In the more westerly parts of north Africa the Jews subsequently achieved considerable power, and there appear to have been large numbers of conversions to Judaism. Many communities, now Muslim in religion, have traditions that they were formerly Jews.

When the racially mixed but Arabic-speaking Muslims invaded north Africa in the seventh century AD, the Jews united with the indigenous Berbers to repel the invaders. The best known story is that of Dahiyah-al-Kahina, Jewish Queen and Priestess, who seems to have combined many of the characteristics of Sisera, of Boudicca (Boadicea), and of Joan of Arc. Commanding a combined army she for years held the invaders at bay until, deserted by her Berber and Christian allies, she fell in a final battle.

When the Muslims in turn invaded Spain in 711 AD and occupied most of the country, many Jews from both north Africa and Babylon followed them and, under the tolerant Muslim regime, there was a great flowering of Jewish scholarship in the new country.

When in 1494, after prolonged warfare, the Muslims were finally expelled from Spain by Christain armies, the Jews expelled with them suffered greatly under the Inquisition, and all who maintained the Jewish faith had to leave the country. Of the expelled Jews, henceforth known as Sephardim, some returned to north Africa and others became dispersed throughout south-east Europe, maintaining the Spanish dialect, Ladino, which they had spoken in Spain.

The Sephardim of Europe have remained a distinct community, but the Jews of north Africa are rather mixed, consisting of members of the ancient Jewish communities, returned Sephardim, and more recent immigrants.

There are in other parts of Africa communities with Jewish customs, notably the Falasha of Ethiopia who do not differ greatly in appearance from other Ethiopian populations, the latter being certainly the result of caucasoid–negroid hybridization very long ago. It is not known how they became judaized.

The Jews of Europe

The origins of the Ashkenazim, the Jews of eastern Europe, are far from clear. We know that Jews were entering Mediterranean Europe in large numbers in the early years of the Christian era. By the third century AD there were already Jews in France, Dalmatia, Scythia, and the Crimea, and considerable numbers in Germany, mainly in the Rhineland. For several hundred years, during the so-called Dark Ages, we almost completely lose sight of the Jews in these countries, though the Sephardim were flourishing in Spain; it is likely that, so long as the power of Rome persisted, and possibly even later, the Jewish colonies in Italy were supplying migrants into France and Germany.

Farther east, we know that the Jews of Persia were crossing the mountains and seas into what is now Soviet central Asia. One very important event was the conversion to Judaism of the rulers, and probably of many of the ruled, of the kingdom of the Khazars, of the Caspian region. The Khazars spoke a language of the Turkish group and apparently had their origins east of the Caspian. Their conversion was well known at the time, and they made no claim of descent from Abraham.

Following the destruction of the Khazar capital, Itil, on the lower Volga in about 965 AD by the Russians, the Khazars gradually disappear from history, though there are sporadic references to them in succeeding centuries as followers of the Jewish religion.

Charlemagne, who reigned from 768 to 814 AD, encouraged the settlement of Jews both in France and the Rhineland. Up to the time of the Crusades the French and Rhineland communities flourished and probably expanded. At the same time Jews were pressing into the more southerly and easterly parts of Germany, apparently from the south. Under the Norman kings they also entered England.

In 1095 AD the Pope summoned Christendom to recover the holy places of Palestine from the hands of the Muslims, and so initiated the long series of wars known as the Crusades. In the course of these the tradition spread that the Jews as a race were guilty of murdering Christ and that it was therefore meritorious to kill them. The massacres which followed, and the Black Death of 1347–8, reduced to a fraction the number of Jews in Europe. At the same time the Jews were expelled from England and France, and in Germany they were chased from one petty principality to another.

Meanwhile, in the middle of the thirteenth century, the rulers of Poland, devastated after the final retreat of the Mongol invaders, encouraged Germans to settle in the cities, and with them came many Jews.

There has been much controversy as to the origin of these Jews, living for the time being in Poland, who were certainly among the main ancestors of the present Ashkenazim. Koestler, in his book *The thirteenth tribe* argues that they were the descendants of the Khazars. Abramsky and others, however, regard them as descended from the earlier Jewish communities of Europe. The evidence is partly historical and partly linguistic and I have tried to sum it up in my book *The genetics of the Jews*. The historical and linguistic evidence points on the whole to a European origin and, as we shall see, blood-group evidence is strongly in favour of the latter.

The Karaite sect of Jews arose in the eighth century AD among the Jews of Babylon. The Karaites are convinced that Jewish religion and practice should be based solely on the ancient scriptures and they reject the more recent interpretative writings, the Mishnah and the Talmud. There were at one time groups of Karaites in a large number of Jewish communities, and modern Karaites do not constitute a single community. There is, however today in Lithuania, in an area formerly part of Poland, a single community of Karaites who hold these beliefs, and who are known to have migrated as a body from the Crimea about the same time as the Ashkenazim entered Poland. They still speak a Turkish dialect, and they claim that they are Jews by religion only, and not by race, a claim that the Nazi authorities accepted during the Second World War, so that they were exempted from the exterminative measures of that regime.

Blood groups

The greater part of our information on Jewish blood groups refers solely to the ABO system. While ABO frequencies are undoubtedly more labile than those of the other systems, we are dealing with events mainly of the last two thousand years, and largely of the last one thousand, so that ABO data still have considerable relevance.

It should be stated at the outset that there is no possibility of finding in Israel any population sample representative of the original pre-Diaspora Jews. The Babylonian Jews from Iraq may possibly give a clue to the genetic composition of these early Jews and it may be meaningful that their ABO gene frequencies lie not far from the centre of the cluster of Oriental Jewish frequencies, especially as the Babylonian Jews differ considerably from the present Arab population of Iraq. These Jews have higher frequencies of both the *A* and *B* genes. Nearly every other Oriental

Jewish community differs in the same manner from its non-Jewish neighbours. The high *A* frequencies and to an even greater degree the high *B* frequencies of the Jews present a general problem. In the context of the high *B* frequencies of the indigenous peoples of Asia, the Jewish frequencies, if due to hybridization, can be explained only by mixing with populations well to the east of their present situations, such as eastern Iran or central Asia, which is on the whole improbable. Frequencies of the blood groups of other systems do not show any systematic differences from those of the surrounding indigenous populations. The Rh groups however need some comment. Some Jewish communities show a raised frequency of the *cDe* haplotype, almost certainly due to hybridization with Africans, but the zero frequency found in two small groups of Kurdish Jews suggests that there was no African admixture in the Jews of the original dispersion under Sargon. However, for one system of blood factors the Kurdish Jews present a special problem. They have, with one possible exception, the highest frequency known of G6PD deficiency, very much greater than is found in the Kurds among whom they live, or lived until recently. Since high frequencies of this deficiency are in most cases certainly a selective response to endemic malarial infection (p. 86), why then do they differ so greatly from the indigenous Kurds? We do not know the answer, but may perhaps cite the example of the Parsis of Bombay who, at a somewhat lower level, have a higher G6PD deficiency frequency than their close neighbours, the indigenous Indians. The explanation in this case is simple. The Parsis, mostly wealthy, had open 'tanks' in the centre of their dwelling compounds, and in these malarial mosquitoes were breeding and maintaining a very local endemicity of malaria.

The Yemenite Jews represent one of the largest ethnic communities in Israel and they have been subjected to a great variety of scientific investigations, including blood-group tests. There is good agreement between the various sets of observations on these, which show that they differ widely from all other Oriental Jewish communities in this respect. They have, in particular, very high frequencies of *O*, *M*, and *S*. In these respects, as in several others, they closely resemble the Yemenite Arabs and other Arabs of the Arabian peninsula. The Arabs, however, have also a very high K frequency which the Jews do not possess. The resemblance is however so close on the whole, and differences from surrounding Jewish and Arab populations so great, as to suggest strongly that the Yemenite Jews and the Yemenite Arabs were not long ago a single community. And this indeed can be explained historically, for the Yemen or Himyar appears to have constituted a kingdom, mainly under Jewish rule, from about AD 200 until 460, when the ruler, Abd Kulalem embraced Christianity, though at least two subsequent rulers followed the Jewish religion. Both Jews and Arabs have rather high frequencies of the African marker genes. Both the northern and the southern Yemenite Jews have such genes, but they are commoner in the north than in the south, and in the Habbanite Jewish isolate frequencies are higher still and comparable with those found in the Arabs.

The advent of Islam in the early years of the seventh century marked the end of Jewish dominance. Many Jews were forcibly converted and many killed, especially in the north of Arabia, but in the south a very considerable body of Jews maintained their faith and became the founders and ancestors of the Yemenite Jewish community which has now migrated to Israel.

The Jews of Africa have consistently higher frequencies both of *A* and of *B* than those of their indigenous neighbours, but the difference is very small indeed in the case of Egypt.

With the exception of the Egyptians the gene frequencies of the non-Jewish population cluster around *A*, 21 per cent, *B*, 12 per cent. The high *A* and *B* frequencies of the Egyptians are well known and possibly ancient.

The Jewish populations, except that of Libya, cluster around *A*, 23 per cent, *B*, 16 per cent, very close to the frequencies found in non-Jewish Egyptians.

It it thus possible that the *A* and *B* frequencies which we now find in north Afrcian Jews are the effect of hybridization with the Egyptians during the sojourn of the Jews in Egypt at the beginning of the Christian era, if not before the Exodus.

The *M* gene frequencies of Jewish communities in north Africa are distinctly higher than the rather low frequencies found in the indigenous populations. As it is generally higher even than the 56 per cent found in indigenous Egyptians, it cannot be explained completely by any incorporation of Egyptian genes.

The chief feature of the Rh groups of north African Jews is the raised frequency, of from 6 to 10 per cent, of the *cDe* haplotype, indicating an appreciable incorporation of African genes.

The Jews of the Tafilalet Oasis in Morocco are certainly an ancient community (which has been dispersed since the blood group survey was carried out). The *B* gene frequency of 29 per cent is exceptionally high, as is that of *M* at 69 per cent. These frequencies are distinctly higher than either in the Berbers to the north or the negroid peoples to the south and if not due to genetic drift must have come from the east. African marker genes are however also present.

The Falasha of Ethiopia and the Lemba of Ẓimbabwe, with many Jewish customs, do not differ greatly in blood-group frequencies from the surrounding indigenous populations.

The Sephardim have been tested mainly on the basis of populations mixed as to their immediate origins, sampled in Israel, but some ABO tests have been done on samples from individual European countries. In all cases *B* gene frequencies, mostly about 16 per cent, are higher than in the indigenous populations, but *A* frequencies are sometimes higher and sometimes lower. Mixed samples have about 24 per cent of *A* genes. The average *M* gene frequency of 56.6 per cent hardly differs either from the south-east European average or from the frequency found in the Babylonian Jews.

Frequencies of from 7.7 to 10.7 per cent of the *cDe* haplotype suggest an African admixture somewhat higher than in the north African Jews, and hence possibly derived from an Egyptian source since the indigenous Lower Egyptians have 19 per cent.

From the blood group data as a whole it may be concluded that the modern Sephardim are descended mainly from those Jews who landed in Spain in the eighth century following in the wake of the Muslim invasions. They probably drew their genes from ancient Jewish communities in Palestine, Babylonia, and Egypt, and from indigenous populations in north Africa. The blood group data suggest that there was relatively little intermarriage with Spaniards.

We possess abundant data on the ABO frequencies of Ashkenazi Jewish populations in Europe, which show that the Jews do not differ greatly in their *A* and *B* frequencies from the indigenous peoples. Frequencies of *B* tend to be somewhat higher in the Jews than in the non-Jews of central Europe, but differences from Russian populations are, on the average, very slight.

MN frequencies differ very little indeed between Jews and non-Jews. In northern and central Europe there is a rather high frequency of the *d* gene in the indigenous populations, averaging about 40 per cent. That of the Ashkenazi Jews averages only 30 per cent, which is slightly lower even than in the Sephardic Jews.

It is rather striking that ABO, MN, and Rh frequencies are closely similar in Ashkenazi and Sephardi Jews. This extends also to the frequencies of the African marker genes and especially the *cDe* haplotype though at 7 per cent it is slightly lower than in the Sephardim. The close overall resemblance leaves little doubt that the Ashkenazim are closely related to the Sephardim, and that both groups of Jews are essentially south-east Mediterranean in their gene frequencies. It is most unlikely from the frequencies found in the Ashkenazim that they could be of Khazar, i.e. south-central-Asian Turkish, origin.

If however we look at some of the non-Ashkenazim of the USSR we see that the Karaites of Lithuania have extremely high *B* and low *A* frequencies, utterly different from those of the orthodox or Ashkenazi Jews of the same region, which closely resemble those of the indigenous Lithuanians.

In the Crimea there are wide differences between the non-Jewish Tartars, the orthodox Jewish Krimchaks, and the Karaites, though all have rather high *B* frequencies. The general conclusion, which I have discussed in greater detail elsewhere, is that the Ashkenazim are related to all the other Jews of presumed Palestinian origin, but that the Karaites of both Lithuania and the Crimea, as well as the Krimchaks, may be descended from the Khazars.

There is a general genetic resemblance between nearly all the other groups of Jews tested, sufficient to show them to have a common origin in the south-eastern Mediterranean area, though modified in varying ways by admixture with the peoples among whom they have lived.

The Gypsies

Another group of peoples who have migrated from Asia into Europe within the last one or two thousand years are the Gypsies. They have undoubtedly come from India; the historical evidence for this is slender but the concordance of the findings of linguistics, social and morphological anthropology, and blood group studies, leaves no doubt as to their origin.

Their precise place or places of origin in India are doubtful, as is the date of their exodus. It is certain that between 1300 and 1500 AD they progressed from eastern to western Europe. They were probably in the Near East about the year 1000, and the most likely reason for their leaving India was the Muslim invasions.

Gypsies have been studied serologically in numerous countries of Europe. So-called Gypsies have also been examined in India, but there are numerous nomad populations in northern India, and it is doubtful which if any of them has a particular relationship to the Gypsies of Europe.

The overall blood-group resemblance to the peoples of northern India is striking, though there are a few exceptions, perhaps to be explained by genetic drift, hybridization with indigenous Europeans, or heterogeneity of origin in India itself. For the ABO system the *B* gene frequency is almost always high. The average *A* and *B* gene frequencies of nearly 5000 European Gypsies are each near 22 per cent, figures which are closely comparable to those found in Pakistan.

As in India, frequencies of *cde* are mostly lower than in Europe and *CDe*, where ascertained, usually higher. The MN frequencies are less consistent. In a small French series the *M* gene frequency is 59 per cent, which is comparable with levels in Pakistan and north India, but a Swedish series shows only 48 per cent and 350 Yugoslavian Gypsies only 43 per cent, both figures being much below average figures both for Europe and for the northern Indian region.

There is a need for considerable further research on this interesting group of populations, with particular reference to those factors which differ in frequency between Europe and the north Indian region.

7. The Pacific islanders

In this chapter the term 'Pacific islanders' will be given a very broad scope, to include the populations of every island from Indonesia, New Guinea, and Australia eastward. It will exclude only the peoples of islands like Hainan which is Chinese, and the Chinese part of Taiwan, also those of Japan, though this country is, strictly speaking, a group of islands bordering on the Pacific Ocean.

Once early man had entered Asia from Africa he was free to reach every part of mainland Asia and Europe which was not covered by permanent ice. But, as far as we know, he had not yet devised boats, or at any rate boats that could undertake voyages of more than a few miles. Thus, with the sea at its present level Indonesia, and still more New Guinea and Australia, were out of reach. But the sea between the Malay Peninsula and all the islands of western Indonesia is very shallow, as is that between New Guinea and Australia. Thus during glacial episodes, when much of the world's water had been deposited as ice sheets on the land, virtually the whole of western Indonesia must have been readily accessible on foot from the mainland of Asia, and Australia potentially so from New Guinea. Also, the many gaps between the islands of eastern Indonesia were much narrowed and all the critical ones could have been seen across in clear weather. Thus even at the *Homo erectus* stage, when he was as far as we know without boats, man could reach Java, and did so about one million years ago. Progress into Australia involved island-hopping across eastern Indonesia (which had to await the invention of boats) and then the long journey across New Guinea and the then dry land of Torres Strait, which was the only way into Australia without crossing long stretches of ocean.

Man, of modern australoid type, had already reached south-eastern Australia by about 30 000 BC, so he must have begun his progress from western Indonesia towards New Guinea a long time, probably several thousand years, before that. We must not think that there was any set plan in his progress; the occupying of each successive tract of new land was an end in itself, and the land had to be suitable for supporting men at the hunting and gathering stage. A group of peoples would probably stay in a newly occupied area for several generations before the need for more food stimulated part of the group to move on.

As we have seen, it was only during glacial episodes that sea level would have been low enough to allow men to cross the present sea areas on foot or by short boat journeys and so to reach New Guinea and Australia. Since, however, these areas were in the tropics they would have been habitable even at the height of such an episode, though rather cooler than at present. Though the most obviously australoid people in Asia now live in the south of India, the deep ocean between India and Australia means that the direct ancestors of the Australians could not have set out from there. We must picture both India and south-east Asia as being at one time inhabited largely by Australoids who were then driven by technically more advanced people from the north, in the one instance into southern India and Sri Lanka, and in the other across Burma and Malaysia and so ultimately through Indonesia and New Guinea to Australia. There are now no clearly recognizable Australoids left on that route. The Indonesians, as we shall see, are relatively recent comers from the mainland of Asia, but the inland peoples of the island of New Guinea are indeed ancient. We shall discuss later (p. 103) their probable relationship to the Australians.

The Australians are a human stock distinct from the Mongoloids, perhaps more like heavily built Caucasoids, with their curly hair, and heavy beards in the males. There are variations in physical type from one part of Australia to another, and attempts have been made to classify them into sub-races, with different origins, but these are a matter for the physical anthropologist; apart from a general difference between north and south, they do not stand out in the results of blood-group tests. Considerable numbers of the latter have been done, but individual series are mostly small, many do not specify tribal names, and for some only overall gene frequencies have been published. In addition there has been considerable hybridization with Europeans. Thus most of the statements made about the blood groups must be generalized ones. It is clear from the facts of geography that aboriginal settlers must have entered from the north and, in so far as successive immigrant populations differed genetically, north–south clines of gene frequencies would be expected, but such clines can be demonstrated almost solely for the ABO system.

The most striking feature is the total absence of the *B* gene, except for an area around the Gulf of Carpentaria where it has obviously come in by sea in comparately recent times.

It is only in the last few years that it has been realized, as a result of archaeological excavations, that man has been in Australia, not for a mere ten thousand years or so as had been supposed, but for at least thirty thousand years, and during most of that time there was probably little movement in or out, so that Australia as a whole was one vast isolate — vast in area, that is, but probably never with a population of more than about 300 000. There are great deserts, but not in the absolute sense of parts of Arabia; there are no very high mountains and no large river systems difficult to penetrate. Thus man, in the long time scale that we can now envisage, could move relatively freely throughout the whole continent, and such processes as genetic drift and natural selection are likely to have affected the population as a whole. One process which especially demands a long time scale is linkage equilibration (pp. 129–30) in such closely linked systems as Rh.

There is indeed a very striking variation in frequencies of the *A* and *O* genes, as we might expect if the ABO system is rapidly and locally responsive to selective pressures, as it appears to be elsewhere. Frequencies of *O* are high in the north — perhaps a climatic latitude effect — while in the south *A* frequencies are high. The appearance of two separate high *A* areas is probably genuine, but it could be the result of lack of data, and random error in small population samples.

The MN system shows high frequencies of *N*, everywhere exceeding 65 per cent, and reaching 95 per cent in parts of Western Australia. The average frequency of about 70 per cent is one of the highest in the world, but not as high as in New Guinea. The *S* gene appears to have been totally absent before the coming of the white man, a unique feature for such a large population group, and one which differentiates the Australians sharply from the peoples of New Guinea. The fact that only one gene occurs at the *Ss* locus deprives us of an opportunity for studying possible crossing-over and linkage equilibration in this system.

The distribution of the genes of the Rh system is also unique. The *CDe* haplotype has everywhere a frequency of over 50 per cent, but this is considerably lower than in New Guinea and Indonesia. The *cDE* haplotype has high frequencies varying from 5 to 44 per cent. The *CDE* haplotype is commoner than in any major group of populations other than the American Indians, locally reaching a frequency of 23 per cent. This unique feature is probably the result of crossing-over (linkage equilibration) between *CDe* and *cDE* in the long period which we now know to have been available for the process and, had we been clever enough, it might have led us to suspect, before the archaeologists proved this, that man had been in Australia for a very long time. The high *CDE* is also important evidence for a relationship to the 'Veddoids' of India. The same process of equilibration probably accounts in part for the frequencies of *cDe*, which are rather high for a non-African population.

The Fy^a antigen has been present in all persons tested — showing a resemblance to the peoples of New Guinea and of east Asia generally.

Tests for the *Gc* plasma proteins, which are carriers of vitamin D, have also been done. The implications of the frequencies of the major genes Gc^1 and Gc^2 are discussed elsewhere (p. 123) but there is also an average frequency of 3 per cent of a special allele Gc^{Ab}, also found in New Guinea. Tests for a number of other blood-group antigens and other genetic factors have been done on small numbers of people but do not at present yield any important anthropological information.

The next large group of peoples to move from south-east Asia into Indonesia, after the ancestors of the Australians, appears to have been the ancestors of the present inland peoples of New Guinea.

New Guinea is geographically very dissimilar to Australia. Though very large for an island it is much smaller than Australia, but much more fertile, a land of very high mountains and deep valleys, now supporting a large indigenous population divided into a multiplicity of tribes with different and not closely related languages and, up to the very recent time of European contact, constantly waging petty war with one another. The wide sea gap which, since the post-glacial rise of sea level, has separated Australia from New Guinea has meant that there has been during this period very little contact or hybridization between the two populations. Though it is possible to find individuals in New Guinea who look like Australians, there is on the whole little resemblance between them.

The archaeology of New Guinea is still in its infancy, but we can be sure that there is a rich harvest to be reaped. Even though we have at present no deposits known to be older than the oldest in Australia, we can be sure that anything datable in Australia, will some day be matched by something even older in New Guinea. The oldest known and dated deposits there, however, go back only to about 25000 years ago.

The populations of New Guinea and many of the adjacent islands are commonly classed together as Melanesians, but this conceals an important dichotomy. Throughout most of the interior of New Guinea we find peoples speaking a great variety of largely unclassified languages, and practising an essentially neolithic way of life. Similar languages are found to a limited extent on the other islands, but the inhabitants of most of these, and of most of the coasts of New Guinea, speak languages of the Austronesian family, certainly much more recently evolved, and spoken by the peoples of most of the oceanic islands. As we shall see when we come to the blood groups, there is a corresponding dichotomy of genetic types. Physically the people of New Guinea, and especially of the interior, are greatly varied in type, but they are dark skinned, woolly haired (and the men bearded), of short stature with strong features, often very prominent noses, which have sometimes been called 'Semitic'. As already mentioned, they show little resemblance to the Australians, though a resemblance has been claimed to the extinct Tasmanians.

The inhabitants of New Guinea have most unusual or extreme frequencies of a number of blood group genes, which are most marked in the peoples of the interior. Detailed work in the north-eastern region has shown that the more extreme features occur in populations speaking non-Austronesian languages, i.e. the languages of the old-established indigenous peoples, who are almost certainly the relatively unmixed descendants of the earliest settlers.

A and *B* frequencies tend to be high in the interior, but the pattern is not a consistent one, and may, like that of the frequencies of these genes elsewhere, be a relatively recent development.

Throughout nearly the whole of the island the frequency of the *N* gene is over 90 per cent, which is the highest frequency known in any large group of populations anywhere in the world. The frequency of the same gene is quite unusually high, over 70 per cent, in nearly all parts of Melanesia; the latter figure is comparable with the rather high frequencies found in Australia. On the other hand the *S* gene, absent among Australians, is present at a moderate to low frequency in most populations, though apparently absent in some. The MN frequencies are perhaps the most revealing features of the relationship between the Melanesians and neighbouring population groups.

Another feature in which New Guinea is unique is in the very high frequencies of the *CDe* haplotype. This has rather high frequencies in Indonesia, in Melanesia, and in much of Australia, but in most of New Guinea it is higher still, and above 90 per cent.

In the Gerbich blood group system the amorph (inactive) allele *Gb* is extremely rare or totally absent in almost all populations outside Melanesia, but it is present in many New Guinea populations, locally reaching a frequency of 80 per cent in New Guinea itself and 18 per cent in the island of Manus to the north. Two hereditary red cell enzyme variants are peculiar to New Guinea, or nearly so. One is a variant of malate dehydrogenase, migrating faster than the normal type on electrophoresis, and the other a variant of phosphoglycerate kinase, which has an X-linked inheritance. In addition, in the Gc system of plasma proteins the allele Gc^{Abo} is found. This is present also in the aboriginal population of Australia but in few other populations.

Thus the old population of New Guinea is a most unusual one, with several unique features, and with a few shared mainly with the Australians. This is consistent with the ancestors of the New Guinea population having followed those of the Australians through Indonesia and part way at least through New Guinea, and having incorporated small groups left behind by the Australians in their migration.

Outside New Guinea itself non-Austronesian languages, apparently related to those of New Guinea, are present in a few of the Melanesian islands. Blood-group characteristics are on the whole similar to those in New Guinea itself, but less extreme, and they bear some similarity to those of south-east Asia and the rest of the Pacific area.

Indonesia

The present inhabitants of Indonesia speak Austronesian languages which appear to have come in from the east, though the peoples themselves are apparently of Malay origin. We have already seen that the island group falls into two parts, those islands which are separated from Asia and from one another by seas sufficiently shallow to have dried out during the ice age, and those separated from the latter by one or more sea channels too deep to have dried out during any pleistocene recession of the sea. This has affected the distribution not only of human but of animal types, and it was noticed by Alfred Russell Wallace, the co-discoverer with Darwin of evolution by natural selection, that a line can be drawn through Indonesia, between Bali and Lombok, separating mammalian faunas of Asiatic type to the west from those of Australian type, including marsupials, to the east.

Wallace himself noticed a similar distinction among the human populations, and this is shown by the blood groups. The ABO system shows relatively little heterogeneity within Indonesia, though there is a slight eastward fall of *A* and *B* and a rise of *O*. There is however a considerable and steady fall in *M* frequency from over 70 per cent in Malaya to under 10 per cent in New Guinea. In the Rh system, *CDe* has frequencies over 80 per cent throughout most of Indonesia, with little variation. Other blood-group systems contribute little to the classification of the Indonesians. The variation in *M* and *N* frequencies is however very definite. It is probably the result of migration of individuals in both directions between Indonesia and New Guinea rather than of any one large population movement. The Philippine Islanders are heterogeneous, with considerable European admixture. There are ABO data available on a considerable variety of population groups, including the negritos who are of particular interest in relation to other negritos in south-east Asia, and the Pacific area. A great part of the rather scanty data for the other blood-group systems refers only to 'Filipinos' without further specification. Blood-group frequencies are, however, in general similar to those of the Indonesians. The tribal populations of Taiwan also speak Austronesian languages and the relatively few blood-group tests done show a general similarity to the Indonesians.

The Micronesians

Micronesia is a very large area of ocean north and east of Melanesia, with a large number of very small islands. The Micronesians speak languages of the Austronesian family. They are undoubtedly related physically and culturally to the Melanesians as well as to the Polynesians, but the people themselves have been less fully studied than either of the groups just mentioned. Also the archaeology of Micronesia is still at a very early stage, but there are indications that the islands were first settled about 2500 BC. There is no doubt that the Micronesians derived much, both of their culture and their genes, from Melanesia.

The blood-group frequencies of the Micronesians do not differ greatly from those of the Melanesians. As in the latter, the ABO frequencies cover a fairly wide range. The *A* gene fluctuates around 20 per cent and the *B* gene averages 13 per cent, much as in western New Guinea. *M* gene frequencies are again near 30 per cent, and *MS* rare or absent; *NS* is present but mostly below 5 per cent. The frequency of the *CDe* haplotype, around 80 per cent, is on the whole lower than in Melanesia, and *cDE* correspondingly higher.

The Polynesians

Polynesia is generally regarded as an island-studded area of ocean with its corners at Hawaii, Easter Island, and New Zealand. It was certainly the last area of Pacific Islands to be inhabited by man. The origin of the people and their culture have long been something of a mystery, and in the popular mind, a rather romantic mystery. The general opinion of anthropologists and of linguists has been that both their language and their genes came from the Melanesia–Micronesia area and hence, in the long run, and perhaps at several removes, from south-east Asia. Thor Heyerdahl, however, advanced the theory that their ancestors came from South America. He subsequently claimed that some came from North and some from South America. Whatever may be their physical ancestry, it appears certain that the Polynesians received their yam and their cotton, directly or indirectly, from America but their pig from the East Indies.

Their languages, all closely related, certainly belong to the same Austronesian family as those of the island and coastal Melanesians, and the Micronesians. It is thought that Proto-Polynesian became distinct from all other Austronesian languages about 3000 years ago. The Austronesian language family appears to have developed and diversified mainly in the small islands of Melanesia many thousands of years ago.

Archaeology shows that it is only since about 500 AD that the Polynesians (or any inhabitants at all) have moved into the present Polynesian islands, but a study of the previous development of various cultural elements seems to show the presence of a recognizably pre-Polynesian culture in Fiji, about 1500 BC.

The physical type of the Polynesians, and particularly their pale skins, differ widely from those of the Melanesians and Micronesians, though containing components which clearly relate them to the latter. Though there appears to be developing a consensus of anthropological opinion deriving the Polynesians and their culture solely from the Melanesian—Micronesian area, there is still considerable uncertanty which makes the evidence of the blood groups particularly important.

The Polynesians differ widely in their blood-group frequencies from the Micronesians and still more so from the Melanesians. Apart from the inhabitants of some of the islands on the western margin of Polynesia, the *B* gene shows very low frequencies and is sometimes absent. It may well have been totally absent before the mixing which followed the arrival of the white man (mixing not only with Europeans themselves but with natives of other island groups). The Polynesians have also an unusually high frequency of the *A* gene, generally above 30 per cent and exceeding 40 per cent in the Cook Islands and Easter Island.

The MNSs system is particularly interesting. The *M* gene, though its frequency is only about 50 per cent, is definitely more common than in the Melanesians and Micronesians. The *S* gene is present but, though its frequency is only 6 per cent, it is present mainly in the haplotypic combination *NS*, the *M* gene being almost entirely in the combination *Ms*. Such a relatively high frequency of *NS* compared with *MS* is extremely rare and (as in the quite separate case of the Japanese, already mentioned, p. 63) indicates a state of linkage disequilibrium between the *MN* and the *Ss* loci. It suggests that two ancient stocks may have contributed to Polynesian ancestry, one having high *N* with a substantial frequency of *NS*. If the combining stocks were both old-established ones each would, in the course of some thousands of years, have reached a state of linkage equilibrium by itself, but the new mixture would be initially, as we indeed find it, in an unstable state. Gradually it would be expected to approach a new state of linkage equilibrium but this would take several thousand years, during which the *S* gene would gradually become evenly distributed between linkages with *M* and with *N*. A high *N* stock, with a moderate frequency of *NS* but very little *M* and virtually no *MS*, could readily have come from a source akin to the present Melanesians; this strengthens the evidence for the participation, in the mixture, of a separate high *M* stock, with little or no *MS*. I had previously used the fact that the Polynesians had more *M* than the Melanesians and Micronesians as suggestive evidence for a high *M* component in their ancestry. The introduction of the linkage disequilibrium hypothesis strengthens the evidence by demanding a high *M* component consisting mostly of the haplotype *Ms*.

The most striking blood-group characteristic of the Polynesians is however the high frequency of the Rh haplotype *cDE*. According to Simmons the mean frequencies of *CDe* and *cDE* are 45 and 54 per cent respectively. Apart from the Polynesians, such high frequencies of *cDE* are almost entirely confined to Amerindians and Eskimos. The general low frequency of *CDE* suggests that, unlike the Australians and Amerindians, the Polynesians have not had time to reach linkage equilibrium for the Rh system.

The frequency of the Fy^a gene is about 80 per cent, and this, though rather high, is much less than the 99–100 per cent found in Melanesians and Micronesians.

The frequency of the Secretor gene is found to be only about 40 per cent and this observation is supported by the high frequency of the Le (a+) blood type, since all Le (a+) persons are non-secretors.

Thus in summary, the Polynesians, despite their wide geographical dispersion, show certain genetic characters which unite the populations of the various islands, and most of which also distinguish them from their neighbours, the Melanesians and Micronesians. The most important of these are the almost complete absence of *B* and the high frequency of *cDE*, both of which also characterize Amerindian populations. The Polynesians also have a moderately high frequency of the *M* gene together with evidence that this comes from a source with a high *Ms* frequency. Amerindians, as well as Eskimos, consistently have high *M* frequencies and in some though not most cases this goes with a high *s* frequency.

The Polynesians also have a lower Fy^a frequency than their neighbours and in this too they resemble the Amerindians. On the other hand they have a low frequency of the Secretor gene, of which the American Indians have nearly 100 per cent. Also the Di^a gene, present in most Amerindian populations, though at very varied frequencies, and absent in a few, is entirely lacking in the Polynesians.

Though I regard it as unlikely that the ancestors of the Polynesians included any substantial number of Amerindians, I think we must give serious consideration to the possibility that these ancestors did include a considerable number of persons originating in the same part of Asia (probably north-east) as the ancestors of the Amerindians and Eskimos.

It is probable, as Captan Cook was the first to suggest, that many of the long-distance voyages, which scattered the Polynesians through the Pacific, were the result of storms which blew adrift many small boats, each with a few passengers. It is thus likely that there have been marked founder effects on gene frequencies. However, these would have affected each island population differently, whereas we are concerned with the consistent way in which the Polynesians differ from their neighbours. If founder effects were involved here, they must have arisen in their original common home, before they were dispersed to their present islands.

8. Amerindians and Eskimos

The last great land area of the world to become inhabited by man was almost certainly the double continent of America. Much more than in Australia, where recent research has pushed back the arrival of man by many thousands of years, the dating of the arrival of man in America and his progress through the continent have provoked very great controversy, chiefly among archaeologists. Did man enter America as long ago as 30 000 or even 40 000 years, or as recently as 15 000 years ago? It is at least fairly certain that he reached the southern tip of the continent some 8000 years ago.

Whatever the date, there is now very little disagreement with the view that he entered across the Bering Strait which was then dry land owing to the recession of sea level which, as we have seen, accompanied the last glaciation. We should however remember that the people who reached New Guinea, and ultimately Australia, at least 30 000 years ago and so possibly before man entered America, must have used boats.

Whatever may have been the precise date of man's entry into America he was certainly then still at the palaeolithic stage of culture. The great neolithic inventions—domestication of animals, agriculture, pottery, were still to come, as was the use of metals. Yet when Europeans first entered America in 1492 all the native populations had reached the neolithic stage and many were using metals. This raises a second major problem of American prehistory, that of the way in which American man discovered these arts. Did palaeolithic man in the Old World and in America discover them independently, or did they reach America across the Pacific from Asia? If the latter was the case, were the secrets brought by a very small number of persons who would not have disturbed appreciably the Amerindian blood group pattern, or was there an important population influx which might have had an effect on blood-group distribution still detectable at the present time? However, any such immigrant group is likely to have been relatively small compared with the indigenous population.

The Eskimos are a group of peoples quite distinct from the Amerindians, and appear to have entered America by sea across the Bering Strait at a late date—probably about 10 000 years ago. They will be described at the end of this chapter.

When Columbus 'discovered' America in 1492 the continent was peopled by a very large number of tribes, speaking correspondingly numerous languages. The study of the very considerable number of languages that have survived, as well as that of the archaeology and history of the tribes themselves, is a highly specialized one. Those readers who wish to study detailed relations between blood groups and anthropology are referred to the author's larger book (Mourant *et al.* 1976) where all known data on the blood groups of the Amerinds (up to about 1970) are tabulated, with tribal names in all cases where they have been recorded.

In this book we shall consider blood group distribution on a broad geographical basis, mentioning the names of populations only where these have a clear relevance to such distribution.

As in the other continents, data on the ABO system are much more abundant than those for the other systems, and of the latter only the MNSs and Rh frequencies have been recorded for a sufficient number of populations to yield clear distribution patterns.

Like the Australian aborigines, the Amerinds appear to have possessed only the *A* and *O* genes before the coming of the Europeans. South of the USA-Mexican border only *O* appears to have been present. Most if not all the *B* genes in the northern zone and the *A* and *B* in the south can be accounted for by interbreeding with post-Columbian immigrant populations from across the Atlantic, but it is possible that some of the *A* and *B* genes in Andean populations are due to trans-Pacific immigration.

Like the absence of *B* in the Australian aborigines, the lack of *B* in the northern zone and of *A* and *B* in the southern zone raises a problem of world-wide importance. Was the *B* gene totally absent from the original populations from eastern Asia that ultimately reached Australia and America, or was the gene lost on the way? If so, was this due to genetic drift in relatively small isolated populations, or to natural selection? Early blood-group workers suggested that when man left Asia for Australia and America mutations for the *A* and *B* genes had not yet occurred. However, analogous if not identical genes occur in the higher apes at least, and so are several million years old. In the light of the discussion of *O* frequencies in Europe it is not difficult to see how, as a result of the elimination of A and B fetuses of O mothers, first the gene *B* (which is rarer than *A*) could have tended to disappear, and then *A* itself. The effects of this form of natural selection could, as has been suggested, have been counteracted by the effects of diseases tending to affect A or B people preferentially. It has already been suggested that such diseases are of an infective nature, and occur in populations which are near together and liable to transmit infections, and not in isolated populations. The populations making their way to Australia and America would have been relatively small, and were progressing, however slowly, into territories without microrganisms specially affecting man, and so would have been particularly free from this hypothetical group of diseases.

This process of unopposed selection could well account for the populations of southern America being without *A* or *B*, and for the loss of B but not A in those further north, but it is still necessary to account for the markedly uneven distribution of *A* in the latter region. Most of the Indian tribes of Canada have high *A* frequencies, especially those of the west, where the Blackfoot and Blood Indians have some of the highest *A* gene frequencies known. Most series of tests on these tribes show over 50 per cent of *A* genes. Both natural selection and genetic drift may have taken part in determining the pattern.

North America

The Indians possessing the *A* gene, living north of a line near the USA–Mexican border, have certain other genetic characteristics in common but not all exclusive to themselves. They have high frequencies of *M*, in fact the gene frequency of 93 per cent in two populations is the highest known anywhere in the world, and the frequency of *S*, mostly as *MS*, is also unusually high. There are few marked variations in *M* frequency within the region but the highest frequencies lie in a broad east–west band in southern to middle Canada. In the Rh system it is likely that *cde* was absent before the coming of the white man. Both *CDe* and cDE have high frequencies, and *cDe* and *CDE* are present, probably resulting from long-term crossing over between *CDe* and *cDE*, as we have suggested in the similar case of Australia (p. 101). The *cDE* haplotype is particularly common, reaching frequencies near 80 per cent in some populations.

The Di^a mongoloid marker gene is present in most populations, but at frequencies of only a few per cent, and generally lower than in the Mongoloids of Asia. The ABH secretor gene, and the PTC taster gene, both have much higher frequencies than in Europeans.

Middle America

A line very close to the southern boundary of the United States marks an important change in several features of the Amerindian communities. This is partly explained by their recent history. North of the boundary they were killed in large numbers, and the survivors segregated in 're-serves'. Thus, though by no means always genetically unmixed, they have remained a totally distinct race. To the south, in Mexico, there was, after the early wars of conquest, less extensive slaughter, so that the Indians have remained very numerous, but they were forcibly Christianized and have in general intermarried extensively with their Spanish conquerors, though many tribes in the less accessible areas have remained almost unmixed.

But not all the existing differences are to be explained by recent history. It was only in Mexico and to the south of it that high civilizations developed before the arrival of the Europeans. There are also genetic differences from the more northerly tribes which are not to be explained by recent history. Firstly there is the sudden disappearance, already mentioned, of the *A* gene. At about the same line the frequency of the Di^a gene rises considerably from the 3 or 4 per cent found in Asia and among the Amerindian tribes to the north. In the present region Di^a frequencies vary widely and rather erratically from zero to 42 per cent, with an average of 9 per cent. The frequency of *M* remains high, though variable, and with no clearly marked pattern, the average *M* frequency being about 75 per cent. The frequencies of *S* are also high and also vary widely, the average being 33 per cent, mostly of course as *MS*.

Rh frequencies do not differ greatly from those found further north, but *CDe* is a little higher and *cDE* though still high is a little lower, averaging about 35 per cent.

In two Nicaraguan tribes the V antigen occurs in phenotypes which seem to show that it is in the haplotypic combination *CDeV*. This, if confirmed, would suggest a relationship, not with Africans, but one, obviously very remote, with the Ainu of Japan. The ABH secretor gene maintains its very high frequency.

South America

There have, in recent years, been very considerable numbers of blood-group surveys carried out on South American Indians by a variety of American and European research teams, but the results have not yet been fully assimilated. We have already discussed the possible causes of the almost complete absence of the *A* and *B* genes, and also of the appearance of low frequencies of these genes in some Andean populations.

The frequency of *M* is mostly high, exceeding 90 per cent in the high Andean region and decreasing gradually from this in all directions. It is lowest on the east coast but it never falls below 55 per cent. The general distribution pattern could almost be an altitudinal one. Frequencies of the *S* gene are also high but fluctuate considerably (as in Middle America) from 5 to 63 per cent, with an average of 30 per cent. Frequencies of the *MS* haplotype are among the highest in the world and are, in particular, higher than anywhere in Asia.

In the Rh system, frequencies of *cDE* continue to be high, averaging about 40 per cent but reaching 80 per cent locally. The *CDE* haplotype has the very high average frequency of 8 per cent. The distribution pattern of *cDE* is somewhat similar to that of *M*.

Frequencies of Di^a are higher among South American Indians than anywhere else in the world, the highest recorded frequency in any sample of over 100 persons being 40 per cent in the Shipibo of Peru. In nearly every country there is a wide range of frequencies which seem to be related to the origins of the populations concerned. In Venezuela the Warao, an isolated tribe of primitive culture, have in one area no Di^a genes at all, and in another, only 2 per cent, probably introduced by interbreeding with Caribs. The Warao are regarded as having entered the region before the arrival of the Diego gene. This may be true of several other tribes in South America who at present show absence or a very low frequency of the gene.

The Indians of South America appear to have, like other Amerindians, a high frequency of the ABH secretor gene, but the data are not very satisfactory. They also appear to have a high frequency of the PTC taster gene. Frequencies of the Gc^1 gene of the Gc system of plasma proteins carrying vitamin D are variable but on the whole high.

The Eskimos

The Eskimos, as already mentioned, entered America from Siberia only some ten thousand years ago. By this time the Bering Strait existed and they had to come by boat. From their landing place in Alaska they gradually spread to the whole of northern Canada and thence to Greenland. A small population of Eskimos still exists in Siberia. A distinct branch of the Eskimo group, the Aleuts, lives in the Alaska Peninsula and the Aleutian Islands, and also in the Komandorskiye Islands off the Siberian coast.

The most conspicuous difference between the Eskimos and the Amerindians is that the former have the *B* gene. This gene is concentrated mainly in Alaska, but is present in nearly every population tested, including the Siberian Eskimos and the Aleuts. The 'pure' Eskimos of Thule, now the most northerly population in Greenland, have no *B*. The *A* gene is present in all Eskimo populations and mostly with a fairly high frequency, between 25 and 30 per cent, and thus higher than in any but a few Amerindian populations.

The MNSs and Rh frequencies of the Eskimos are closely similar to those of the Amerindians. The frequency of the *M* gene is only moderately high in Alaskan Eskimos and Aleuts, with 58 and 64 per cent respectively, but most Eskimo populations have over 70 per cent of *M*, and frequencies exceed 80 per cent at Point Barrow in Alaska, and in several populations in Greenland. Frequencies of the *S* gene are mostly near 20 per cent, and the gene is found mainly in the combination *MS*.

The *d* gene is probably totally absent in unmixed Eskimos. Frequencies of *cDE* are high, mostly just above 50 per cent in the Eskimos of Alaska and Hudson Bay but somewhat lower in the Aleuts, the Eskimos of Kodiak Island (Alaska), Southampton Island (Canada), and Greenland.

Another difference from the Amerindians is the complete or almost complete absence of Di^a, but it must be remembered that the latter is rare in the North American Indians. Incidentally the Polynesians who share with the Amerindians and Eskimos a high frequency of *cDE*, also lack Di_a. The frequency of the Gc^2 gene, for which there are abundant data, averages about 30 per cent, which is higher than in most North American Indian tribes. This may, as suggested elsewhere (p. 123), be the result of climatic selection.

The abundant data both on Eskimos and Amerindians for the ABO, MNSs, and Rh systems prompt a comparison of frequencies in the two major populations.

Amerindians and Eskimos

The close resemblance between the Eskimos as a whole and the Amerindians as a whole, in their MNSs and Rh frequencies, suggests that they are drawn from the same stock despite the long interval between their respective arrivals in America. This conclusion is strongly supported by very recently published work on variants of several plasma proteins and red cell enzymes in the Indians and Eskimos of Alaska. Taking all these results together there is now virtually no room for the alternative hypothesis of convergent evolution, which indeed is most unlikely in view of the very wide range of environmental conditions to which these peoples have been exposed since their arrival in America.

There is still a need for assimilation of the new results, and for further studies on similar lines. It is unfortunate that there are still important gaps in our knowledge of the frequencies of the blood groups (in the restricted sense) of key populations, especially since the new data concern genetic systems for which data are still rather scanty outside Alaska. A comparatively few gap-filling tests on selected populations would make it possible to use the very large body of existing observations much more efficiently in working out the relations between the various peoples. Already the available results, genetic, linguistic, and archaeological, have given us a much clarified picture of the arrivals, migrations, and divergences of the Amerindians, the Aleuts, and the Eskimos.

Dr Harper concludes that what he calls the Bering Sea Mongoloids split off from the Siberian Mongoloids 19 000 years ago, that the Amerindians, or at least the Athabascan section of these, separated from the Eskimos as a whole 15 000 years ago, and that the Eskimos in the narrow sense separated from the Aleuts 9000 years ago. There will certainly be continuing controversy as to the validity of these dates but the essential framework is now available. As shown by Harper, there is now also the beginning of a conceptual bridge connecting Alaskan populations archaeologically and genetically with those of Siberia.

One problem which needs further consideration is that of the ABO groups, in view of the absence of the *B* gene from the Amerindians, and of the *A* gene as well from those of the south, whereas both genes are present in the Eskimos and in all known Siberian peoples. If the explanation given above (p. 109) of the absence of these genes in Amerindians is accepted, it would not be surprising if the relatives remaining in Asia, like the late-arriving Eskimos, had retained these genes.

9. Causes of gene frequency change

Every population, as we have seen, is characterized by a set of gene frequencies for each blood group system. In the short term, over a period of a few generations, these frequencies change very little, so that they can be used to identify a population which has migrated, as for instance Scots or Africans to America, or Germans to Russia. Yet, in the long term, frequencies have certainly changed, and populations which undoubtedly have a common origin but have been separated for hundreds of generations are found to have widely different frequencies. There are four main causes of gene frequency change, mutation, natural selection, genetic drift, and linkage equilibration; these were briefly described in Chapter 2 (pp. 18–19). We must now consider these causes, and especially natural selection, in greater detail.

Mutation

Mutation is constantly taking place in every genetic system, but for any one system it is an extremely rare event, occurring only about once in a million reproductions. Its main effect is to provide the initial material upon which natural selection and genetic drift operate to produce the large frequency variations which we actually observe. Blood-group gene mutations have been observed directly only in a handful of instances.

Genetic drift

It is a matter of observation that in small isolated populations the frequencies of many of the blood group genes differ widely from those found in the larger populations from which they are descended. The cause of isolation may be physical, as in the populations of islands, or valleys surrounded by high mountains, or they may be religious, as in the Samaritans of Palestine, or various strict Protestant sects in America. We have already considered the possible operation of genetic drift, probably interacting with natural selection, in determining the blood group frequencies of the Icelanders (pp. 79–80).

There is general agreement upon drift as a cause of aberrant gene frequencies in such small populations. However, drift should be taking place, even if extremely slowly, in populations of any size, and some authorities regard it as a major long-term cause of many of the observed frequency differences between large populations, especially in systems where little evidence has been found of the operation of natural selection. Indeed, some investigators regard natural selection as having played only a small part, if any, in causing the present distribution of the blood groups—which leaves genetic drift as the only candidate for the main cause. There are however numerous cases where natural selection appears to me to have had or to be having an important modifying influence on the frequencies of blood groups and other hereditary blood factors, and most of this chapter will be devoted to setting out the evidence for this process.

Natural selection

There is no doubt that man's anatomical and physiological characters are adapted to his environment both in the world as a whole, and in different parts of the world. However, because not only biological characters, but also cultural features, are passed on from parents to children, it is often difficult to disentangle their effects.

The blood groups and other inherited biochemical characters are of course determined purely genetically, so that there is no question of any cultural influence on their expression. In the short run they are determined solely by the genes received from the previous generation. The problem here is how far the frequency changes which have undoubtedly taken place are the result of chance genetic drift, and how far the result of natural selection. And, if they are due to selection, what are the precise features of the environment which in any particular case are causing the selection. We must remember that not only features of the external environment, such as climate and prevailing infections, may cause selection, but also, as we shall see, the very combinations of genes in the population itself may have selective effects on the frequencies of particular genes.

In a given blood-group system the differences between individuals concern certain substances on the surface of the red blood cells. Relatively large molecules are firmly anchored in the surface membrane of the red cells, but have attached to them, and perhaps waving like seaweed in the surrounding sea of plasma, short molecular chains consisting usually of simple sugars, differing slightly but specifically according to the particular blood group of a given system. In some systems the specific side-chains consist of amino-acids. Many of these chains are similar to those found on the surface of bacteria, and the serological tests used for distinguishing between the blood groups are analogous to certain tests used in distinguishing between strains of bacteria. It might therefore be expected that blood group differences would in some way control susceptibility to infection by particular strains of bacteria. Since response to infection is also partly determined by environmental conditions such as atmospheric temperature, the environment might participate in selectively affecting blood group frequencies.

There are two main lines of approach to the problem of differing susceptibility to particular infective diseases. One is the study of the biochemical and immunological properties of the blood-group substances on the red cells, and comparison with the corresponding properties of bacterial substances. This approach has taught us a great deal about the nature of the blood-group substances, but little about their possible effect on susceptibility to infection.

The main alternative approach is the statistical one, comparing the frequencies of the blood groups of a given system in persons suffering from particular diseases, infectious and other, with the frequencies found in the population as a whole. If the frequency of a given blood group is observed to be higher in sufferers from a particular disease than in the general population, this may be explained in a number of ways. The most likely one is that persons of that blood group are more susceptible to the disease than persons of other blood groups. But before that conclusion is reached certain other possibilities must be eliminated. There may be what is known as a stratification of the population. For instance, if it were found that in Britain the frequency of group B was higher in sufferers from sickle-cell anaemia than were healthy persons, this would almost certainly be because Negroes have a much higher incidence both of sickle-cell anaemia and of group B than the white population, though there may be no direct relationship between the two conditions.

A false appearance of correlation between a blood group and a disease could arise if the blood group frequencies in the control population were wrongly estimated—for instance if they were based on a series of blood donors who were not a true sample of the general population. This might arise because of a tendency to select group O persons as donors. This would give a false appearance of a positive correlation of the disease with groups A and B and a negative one with group O.

The classic example of a correlation between a disease and a hereditary blood character concerns not a blood group but a haemoglobin variant. It will be described in detail later but the broad situation must be set out here in order to show the kind of phenomenon for which we are looking in the blood-group systems.

As already briefly mentioned on pp. 32 and 46, the gene for normal haemoglobin (Haemoglobin A) is partly replaced in many African populations, and in some others, by an allelic gene, for Haemoglobin S or sickle-cell haemoglobin. Homozygotes for S nearly all die in infancy of sickle-cell anaemia, but the gene, despite this apparent selection against it, persists at a high level in many populations. This is now known to be because heterozygotes for Haemoglobins A and S are much more resistant to malignant tertian malaria than are normal persons with only Haemoglobin A. The result is a balance between selection by malaria and selection by sickle-cell anaemia, so that the population settles down to a Haemoglobin S frequency related to the local incidence of this form of malaria.

One reason why this particular example of selection was discovered relatively early is the very definite difference between the haemoglobin genotypes in the severity of the symptoms of malaria from which they suffer. On the other hand a genotypic difference for any one blood-group system represents only one out of numerous sets of influences on the severity of a particular infection or other disease.

Nevertheless certain diseases are much more common in persons of one blood group than of another—or rather, the frequency of a particular blood group is very significantly greater in persons with the disease than in the general population. A desire to clarify the situation led the author and his colleagues to carry out an analysis of all the availble data and to publish a substantial monograph on the subject, and the next few pages are largely based on that analysis.

Because of the immunological resemblances between certain blood-group substances and certain bacteria the emphasis was at first on infectious diseases rather than others, but in the outcome, while important associations were found with certain infections, the most marked, and in general unexpected, associations were with other diseases, such as cancer and thrombosis.

Some associations with ABO groups

Some of the most marked associations with microbial diseases are those with streptococcal infection, which tends to affect persons of group A preferentially. This is not unexpected in view of the biochemical similarities between the blood-group antigens and those of this class of bacteria. A great deal of the data on bacterial infections such as tuberculosis and leprosy are difficult to interpret, but in general it may be said that there is a tendency for bacterial infections to attack persons of group A most severely, while virus infections tend, in a very general way, to be associated with group O.

The most striking associations are however with cancers, nearly all of which are associated with group A, as are clotting diseases, while bleeding diseases, mostly due to a deficient clotting mechanism, are, on the contrary, associated with group O.

Other diseases which appear to be associated with group O are the autoimmune diseases. These are diseases in which the body makes antibodies which tend to destroy certain of its own cells and tissues, one of the best known examples being the destruction of the red blood cells in autoimmune haemolytic anaemia. Other autoimmune conditions include various thyroid diseases, myasthenia gravis, multiple sclerosis, and possibly rheumatoid arthritis. However, the data for this most interesting type of association are at present rather slender and many more observations are needed.

If however the association is a real one, the contrast with the cancer–group-A association is an interesting one in view of the suggestion of Macfarlane Burnet that there is a fundamental antithesis between the two classes of disease. One of the causes of clinical cancer is probably that the body has failed to destroy the cancer cells at an early stage in their development by making antibodies against them; this implies a certain weakness of the immunological system, and would probably be accompanied by a low susceptibility to autoimmune diseases generally. Persons with autoimmune diseases, on the other hand, have overactive immune systems and will tend to destroy not only their own normal tissues but also cancers at an early stage in their development. There are indications that these two classes of disease do tend to be mutually exclusive.

In addition to (or perhaps in parallel with) their statistical association with group A, some cancers contain an A-like substance even when they occur in persons not of group A or AB.

These observations suggest a more general hypothesis, that in the tissues, both normal and neoplastic, of all persons, there are blood-group-A-like antigens present at a biochemical level at which they are usually inaccessible to the immune system. However, in the course of an autoimmune process, or of the immune response to a growing cancer, the antigen may become accessible. Then an A person, who cannot make anti-A, will be more likely than an O person to tolerate the cancer, but less likely than an O person to attack his own tissues. If this is true of A and O it may yet be found to be true also of other pairs or sets of genetically determined antigens.

Ulcers of both the stomach and the duodenum appear to be associated with group O, but it is uncertain how far these apparent associations are due to the bleeding (associated with group O) which tends to bring the patients into hospital, and so into the net of the statistician. While however infectious diseases tend to cause debility and death in early life, and this will tend to inhibit reproduction and so affect the blood-group frequencies of the next generation, the other diseases and conditions mentioned tend to occur in later life, and so will have little effect on reproduction and on the composition of the following generation.

Before we go on to consider the other blood-group systems we must look at a type of selection in which nearly all of them are involved to some degree. This is that due to haemolytic disease of the newborn, and we have already studied the most important example of this, the one due to the Rh blood groups. However, in any blood-group system, including the ABO, a baby will often inherit from its father a blood-group gene, and the corresponding blood-group substance, or antigen, which are not present in the mother. Normally the blood circulations of the mother and the unborn child are separated by a membrane within the placenta, but occasionally this is breached, and some of the red cells of the child pass into the mother's circulation. The child's blood-group antigens may then be treated by the mother's antigen-forming system as foreign intruders, in the same category as infective micro-organisms, and she may manufacture antibodies to destroy these substances and the cells which carry them. The resulting antibodies, unlike the cells, can traverse an intact membrane. Such an 'immunization' process usually occurs only late in pregnancy, too late to affect the child which caused it, but if the mother has another child of the same blood group it will be subject to the destructive process throughout pregnancy, and so be born with haemolytic disease, or even die of it before birth.

The three systems chiefly involved are the ABO, Rh, and Kell. We have already discussed (p. 28) the effects of Rh immunization. In the Kell system it is chiefly the K-positive infants of *kk* mothers who get the disease.

The position with regard to the ABO system is more complicated. It is the A (genotype *AO*) and B (genotype *BO*) infants of O (genotype *OO*) mothers who are sometimes affected. Such mothers, even before becoming pregnant, already (like all group O persons) have both anti-A and anti-B antibodies in their serum, but in a form which does not readily cross the placenta and so does not affect the infant. However, in a small proportion of cases the mother develops an anti-A or anti-B antibody of a somewhat different kind which does traverse the placenta and cause haemolytic disease in the infant. It has also been shown, by large-scale surveys, that quite apart from haemolytic disease of the newborn, which as the name implies, tends to cause disease in full-term newborn infants, the A and B infants of O mothers also have an unusually high tendency to undergo abortion early in pregnancy. It is thought that in such cases the maternal antibody is attacking not the red cells but certain other fetal tissues which carry the A or B antigen.

Since in nearly all populations the frequency of the O gene is over 50 per cent (see p. 75) the ultimate effect of the deaths will be an increase in frequency of *O* and a decrease of both *A* and *B* genes, so that ultimately these two genes should disappear completely. Such total absence is the situation in many American Indian populations, and the process mentioned is perhaps the cause of it, as it may be of the absence of the *B* gene in nearly all Amerindian and Australian aboriginal populations. But *A* and *B* should disappear in all other populations as well, which they do not seem to be doing. We have already seen that *A* and *B* frequencies tend to be low and *O* high in most of the relatively isolated island, mountain and desert populations of Europe, South-west Asia, and North Africa, and the suggestion is here put forward that once again it is haemolytic disease that is responsible. Why then do the *A* and *B* genes persist at fairly high levels in the people of other areas? Can it be that in areas with free communication with the rest of the world, in contrast with isolates, some process is going on which favours the *A* and *B* genes and so maintains a balance? Such a process could be the occurrence of certain epidemic infections from which the-isolated populations are protected, and which tend to occur especially in group O persons. The statistics of associations of epidemic diseases with group O are not sufficiently clear to enable us to dogmatize on this point, but the mechanism is a possible one (see Table 5, p. 80).

The Rh blood groups

Apart from the special case of haemolytic disease of the newborn, the Rh groups show few if any convincing associations with diseases. The most probable one is that of the Rh-positive (D-positive) type with duodenal ulcers.

The MNSs blood groups

There are no definite associations of diseases with particular groups of the MNSs system. There is a single claim for close association of carcinoma of the breast with S-negativeness, i.e. with the genotype *ss*.

The Duffy blood-group system

A great many of the blood-group antigens of nearly all systems have been found to cause rare cases of haemolytic disease of the newborn, but one of the most striking examples of an association with any blood group with a disease concerns the Duffy system and malaria. This discovery was made as a result not of a statistical study but of laboratory work. There are three relatively common alleles in this system, Fy^{a}, Fy^{b}, and Fy^{4}, each producing a characteristic antigen. The genes Fy^{a} and Fy^{b} together account for about 99 per cent of the Duffy genes of non-Negroids, but among the unmixed Negroids of Africa these genes are rare and the gene Fy^{4} has a frequency nearly always above 90 per cent.

It has long been known that Negroes are almost totally resistant to infection with the malaria parasite *Plasmodium vivax* which is highly infectious to all other races, and which causes benign tertian malaria. The common type of malaria in Africa is malignant tertian, caused by the parasite *Plasmodium falciparum*. Consideration of these two apparently unrelated sets of facts led Dr L. H. Miller to experiment with red cells of various Duffy genotypes, trying to infect them with a malaria parasite. For safety's sake he used *Plasmodium knowlesii* which infects monkeys, but is closely related to *P. vivax*. He found that cells of all genotypes involving Fy^a and Fy^b were readily infected, but those of type Fy^4Fy^4 were highly resistant. He concluded that it is the Fy^a or Fy^b antigen molecule on the surface of the red cell that acts as the portal of entry of the *P. vivax* parasite. It will unfortunately be difficult to confirm this conclusion by direct observation on man. Failing the dangerous deliberate infection of human volunteers with malaria, it would be necessary to study a population which carries all three genes Fy^a, Fy^b, and Fy^4 in substantial frequencies, and in which *P. vivax* infection is endemic and relatively uncontrolled. There probably are still such populations in the Near East and in Madagascar.

Dr Miller's theory enables us to reconstruct the history of the response of Africans to malaria as occuring in two stages. *Plasmodium vivax* was formerly endemic in Africa, as it was until the present century in many other parts of the world. Then mutation occurred in the Duffy blood-group system, Fy^a or Fy^b mutating to Fy^4. It was presumably a matter of chance that the new gene spread in some community sufficiently to give rise to a few homozygotes. The latter proved to be almost completely resistant to the prevailing *P. vivax* malaria, and so the new gene spread rapidly by natural selection, first locally and then throughout tropical Africa, a process perhaps taking as much as one thousand years. Then, perhaps for some hundreds of years, tropical Africa was almost totally free from malaria until *Plasmodium vivax* underwent a mutation (or perhaps several successive mutations) converting it into *Plasmodium falciparum* which is both readily able to infect Fy^4 homozygous red cells, and is in general much more virulent than its predecessor. It, in turn, was able by natural selection largely to replace that predecessor. It should be possible, in the history and archaeology of Africa, to see the effects of the resulting sudden continent-wide deterioration in health.

But now man in turn responded genetically to the change. Either by introduction from Asia, or by mutation, the sickle-cell haemoglobin gene entered Africa. Heterozygotes for the new gene were highly resistant to the effects of *P. falciparum* infection, and so the sickle-cell gene in turn spread to large areas of Africa and helped to keep the new virulent type of malaria in check. But this process was a much more costly one than that involving Fy^4, for it entailed the deaths in infancy of nearly all sickle-cell haemoglobin homozygotes.

There are indications, in Sardinia and New Guinea, that a number of other hereditary blood types affect susceptibility to malaria. One of these is thalassaemia, a condition affecting haemoglobin synthesis. Here, as with the sickle-cell condition, the heterozygotes are apparently somewhat more resistant to malaria than are normal non-thalassaemic persons, while the homozygotes tend to die young of anaemia.

The plasma proteins

In addition to blood groups in the strict sense, there are, as we have seen, a very large number of genetic systems, affecting the proteins of the blood plasma, and enzymes in the red cells. Most of the plasma proteins which are subject to hereditary variation have known physiological functions and in many cases their reactivity varies with the genotype.

As an example we may take the Gc groups. The protein concerned was known by its physical and immunological properties as a component of plasma, and it was known to have a number of hereditary variants characterized by different speeds of migration on electrophoresis. But for many years after all this had been discovered, it had no known function. Then it was found that the Gc proteins are the carriers of vitamin D. This vitamin either comes from food or is synthesized through the effect of the ultraviolet part of sunlight acting on a substance in the skin. Vitamin D is essential for the proper absorption of calcium from food, and so for proper bone formation, and a deficiency of it causes the bone disease known as rickets. It is now known that the hereditary variants of the Gc proteins differ in their efficiency as vitamin D carriers, and it is to be expected that the more efficient one will prevail in conditions of poor sunlight. There are indications that this is the case, for Gc^2 is on the whole more common, and Gc^1 less so, in conditions of poor sunlight. The relation is however not a constant one, and the problem is currently under intensive investigation.

Individuals with two low-activity or amorph genes for the protease inhibitor (Pi) system suffer from one or more diseases characterized by protein breakdown. They invariably have some degree of pulmonary emphysema, with breakdown and running together of the alveoli, the little air sacs of the lungs. They also often have a form of liver cirrhosis, and sometimes duodenal ulcers. They are probably also more than normally liable to a number of other diseases.

The protease inhibitor or *Pi* variants certainly vary considerably in frequency in different populations and one might expect that, for instance, low or absent Pi activity would suffer negative selection in populations where pulmonary function is particularly important, such as high altitudes, but so far no such functionally based population studies appear to have been done.

As with so many pathogenic genes, it is problematic how the harmful genes of the Pi system persist with substantial frequencies in most populations. One suggestion is that their harmful effects are balanced by a selective advantage in that the spermatozoa of males with these genes, containing little or no protease inhibitor, can without inhibition break down the proteins present in solution in the fluids of the female reproductive tract, and so penetrate these fluids more easily, and that such males are therefore more fertile than normal.

The secretor system

The *Se* and *se* genes of the ABH secretor system (p. 30) vary widely in frequency in different populations, but our knowledge of their distribution is incomplete and patchy. In particular the *Se* gene has a very high frequency in American Indians and apparently a low one in southern India. The system shows well-defined associations with certain diseases and there are indications that it is involved in major processes of natural selection.

The most important group of diseases so associated is that of the peptic ulcers—gastric and duodenal—which are markedly associated with non-secretion: The association of O blood groups with the same diseases is partly at least a statistical accident, since an important part of it really is due to the association of haemorrhage with group O and, as we have seen, it is the haemorrhage which so often brings the patient into hospital, and into the scope of blood-group surveys. The association with non-secretion is however intrinsic to this class of diseases and is independent of haemorrhage.

There is however another physiological function of the intestinal lining which has important genetic implications. All individuals have in their intestine an enzyme which breaks down organic phosphates—alkaline phosphatase. This particular phosphatase can be recognized by its chemical properties and its speed of electrophoresis, and it has thus been found that some people do and others do not have intestinal alkaline phosphatase in their plasma. Tranfer of this phosphatase from the intestine to the plasma is controlled by the ABO blood group and the secretor systems—it is promoted by the secretor factor and suppressed by the A antigen, whether as group A or AB. This suggests that peptic ulceration is related to the ability or non-ability of the intestinal membrane to let the phosphatase pass into the plasma.

The tendency towards carcinoma of the stomach, markedly associated with group A, may also be affected by these membrane properties, though there seems to be no association between this form of carcinoma and secretion. The subject is however one that demands further research.

By analogy with the selective favouring of group O through the elimination of A and B offspring of group O mothers, there is probably also a selective elimination of secretor fetuses (of groups A and B) of group O (but not necessarily non-secretor) mothers. This results from the fact that fetuses which provoke the immunization of such mothers are more frequently secretors than are members of the general population.

It may thus be that there is a constant selection against the secretor gene as a result of ABO haemolytic disease of the newborn, and that the persistence of the secretor gene at varying levels in different populations is the result of environmental conditions which in varying degree favour secretors, and so more or less counterbalance the selection by haemolytic disease. The situation is by no means as clear-cut as the selection against A and B but merits both studies of cases of ABO haemolytic disease of the newborn, and population studies of the distribution of secretors and non-secretors, as well as further research on possible disease associations of the secretor and non-secretor types.

The main diseases known to be so associated are gastric and duodenal ulcers, and 'arteriosclerosis', all associated with non-secretion, and diabetes mellitus, associated with secretion. Carriers of haemolytic streptococcus show a fairly consistent deficiency of secretors, as does rheumatic fever which is also a sequel to streptococcal infection. Much more attention should be given to the combined effects of blood group and secretor state on susceptibility to bacterial infections.

The haemoglobins

We have already seen that the haemoglobin system is subject to very strong natural selection (pp. 46, 117). The association of one haemoglobin variant with malaria infection has already been discussed briefly. It must now be looked at more closely.

Haemoglobin is a complex protein in the molecule of which there are four chains of amino-acids (linked by peptide bonds) comprising two identical alpha chains, and two identical beta chains. The alpha chains and the beta chains are controlled by separate sets of allelic gases. It is the variants of the beta chain which are of greatest interest both medically and anthropologically. Usually the difference from the normal consists in the change of one out of several hundred amino-acids in the chain. In various parts of the world, but especially in Africa, the gene for the beta chain of normal adult haemoglobin, or Haemoglobin A, is partly replaced by the gene for a variant, Haemoglobin S, or sickle-cell haemoglobin. If we represent the normal gene (for Haemoglobin A) by the letter *A*, and that for sickle-cell haemoglobin by *S*, then in a population having both genes there will be three genotypes present, *AA* (the normal), *AS* giving rise to a mixture of the two haemoglobins, and *SS*, giving only Haemoglobin S. Persons of *AS* genotype are almost perfectly healthy, though their red cells under certain conditions curl up into a sickle-like form, but *SS* persons, who show this sickling phenomenon to an even greater degree, nearly all die in infancy from sickle-cell anaemia because of destruction of the distorted red cells. Thus in each generation a substantial proportion of the *S* genes are eliminated before those possessing them have a chance to reproduce. If this were the whole story then the *S* gene would disappear almost completely in a few generations. However, many African populations have a high and apparently stable frequency of *S* genes. This was for long a puzzle, the first stage in the solution of which was published by A. C. Allison in 1953. It was found that *AS* heterozygotes were more resistant than normal *AA* persons to malaria of the malignant tertian type, caused by the parasite *Plasmodium falciparum*. Thus, in the presence of this parasite, *AA* persons tend to die of malaria, *SS* persons nearly all die of sickle-cell anaemia, but *AS* persons are resistant to both diseases, and thus contribute both *A* and *S* genes to the next generation. Thus the frequency of the haemoglobin S gene is maintained at a level related to the degree of infection of the population with *P. falciparum*.

There are, however, other beta-chain alleles, giving rise to other haemoglobin variants, which have a frequency of several per cent in certain populations.

Haemoglobin E is found throughout south-eastern Asia and north-eastern India, Haemoglobin D with frequencies of about 3 per cent in parts of south-west Asia, especially the Punjab and Gujerat. Haemoglobin C is almost confined to West Africa where it has high frequencies and overlaps with Haemoglobin S.

For none of these is there more than slight debility in the homozygotes. but there is probably a slow selection against the respective genes. Also where any one of them coexists in the same individual with thalassaemia, so doubly depressing the synthesis of normal Haemoglobin A, there is again anaemia. Thus, to account for the maintenance of these genes, we have to look for diseases against which the heterozygotes have some relative protection. For Haemoglobins C and E as for S there is some evidence for protection against malignant tertian malaria, but other diseases also have been suggested as selective agents favouring one or another abnormal haemoglobin.

The histocompatibility antigens

The histocompatibility antigens are very closely associated with a wide range of diseases. The implications of this for evolution and natural selection are enormous, but at present the contribution to anthropological knowledge has not been great—chiefly because only rather small numbers of non-caucasoid people have been systematically surveyed for their histocompatibility antigens, so that it is as yet difficult to see world patterns such as we can see for many of the blood group antigens. It must also be admitted that we have not yet learned to see the wood for the trees. There are so many alleles at each of the four loci, and so vast a number of possible haplotypes composed of these alleles, that it is difficult to fix attention on those which in a few years we shall perhaps see to be most significant. In due course, however, the richness of the information now becoming available seems likely to have a classificatory value, and perhaps an evolutionary importance comparable to that of all the other hereditary blood factors combined.

The most striking association found so far is that between HLA-B27 and ankylosing spondylitis. The antigen has a frequency of 9 per cent in the general British population, but in patients with ankylosing spondylitis the frequency is 90 per cent. This of course implies an association which is statistically very highly significant indeed. But its significance derives mainly from the very close association rather than from the total numbers tested, and it would stand out even if only a very small population sample had been studied. The same is true to some extent of all the histocompatibility associations. The blood-group associations are on the whole much more slender. The frequency of cancer of the stomach is only 21 per cent greater in persons of group A than of group O. The significance (or perhaps we can say the convincingness) of the association depends on the fact that this slightly increased frequency is consistently shown by almost every survey carried out, comprising in all 63 000 patients.

Each class of associations implies natural selection against the phenotype and the gene associated with the disease, but the rate of selection must be much greater for the histocompatibility antigens than for the blood groups.

There are, however, other differences between the kinds of association which we find with the blood groups and those with histocompatibility antigens. The blood-group associations are rather heterogeneous as regards type of disease, and few of them could have been predicted in advance.

The histocompatibility associations are, on the other hand, largely of one particular kind. The diseases are mostly (though by no means all) of kinds which are due to disturbances of the body's immune mechanism, and a great many of them belong to the class of autoimmune diseases, diseases in which the body's antibody-making process is turned against its own tissues and organs. It is known, moreover, that the closely linked chromosome segment controlling the immunoglobulins also includes genes for proteins with immunological functions other than as immunoglobulins.

An association of a particular gene with a disease implies, as we have seen, a selection against that gene. But in many cases the association is not merely with a particular gene, but with a particular haplotype, or combination of genes at two or more of the separate (but closely linked) histocompatibility loci.

In any linked set of genes, even a closely linked one, the tendency is, as we have seen, for genes at one locus to distribute themselves evenly, by crossing-over, between the genes at the other loci. But associations with combinations of genes, or haplotypes, imply that these haplotypes, and not simply individual genes, are being involved as entities in the selective process, so counteracting the tendency of the genes at one locus to get distributed evenly between those at the other loci. Where there is a measurable, even if slow, rate of crossing-over between the loci, such a selective process must operate if linkage disequilibrium is to be maintained. With the even more closely linked systems such as Rh and MNSs, crossing-over will be rare and haplotypes much more stable, so that only a small amount of selection for or against particular haplotypes will be needed to maintain the *status quo*. The general question of stability of linkage equilibrium is further discussed below.

The phenythiocarbamide tasting system

There is one genetic system which has been widely used in population surveys, which is not, as far as is known, in any way a blood factor system. This is the phenylthiocarbamide tasting system (pp. 33–4).

As already mentioned, phenylthiocarbamide is a thyroid inhibitor, and knowledge of this led Professors Harry Harris and H. Kalmus, and Dr W. R. Trotter, and later Dr F. D. Kitchin and his colleagues, to look for an association between tasting and thyroid diseases.

They and others have shown that there is indeed such as association, for persons with ordinary nodular non-toxic goitres include an excess of non-tasters while those with toxic goitres, and overactivity of the thyroid gland, include an excess of tasters. There is some evidence that even in persons who are clinically normal there is a higher frequency of tasters among those with higher thyroid activity, and that tasters tend to develop more rapidly at puberty than non-tasters.

In this system we have a particularly complete picture of the environmental factors involved. As we have seen, the thyroid hormone molecule contains iodine, so that, for normal thyroid activity, a certain level of iodine is needed in the diet. Such small quantities of iodine, normally occurring as iodide, are tasteless to all individuals. It is moreover known that unduly high levels of iodine can produce thyrotoxicosis. On the other hand there are often present in the diet, especially in cabbages, thiocarbamide derivatives which are thyroid inhibitors, and which PTC tasters taste as bitter. We may suppose therefore that tasters, but not-tasters, will limit their intake of such substances, and so be more liable to thyroid overactivity, and less to underactivity, than non-tasters. Since both underactivity and overactivity can affect fertility and can indeed be fatal, we can envisage a delicately balanced polymorphism of the two allelic genes, based on the levels of iodine and of thyroid inhibitors in the diet, such that if iodine is deficient or inhibitors in excess, tasters will be favoured selectively, while if there is an excess of iodine, or a lack of inhibitors, non-tasters will be favoured.

Sunderland has drawn attention to a particular case where differences in taster frequency are possibly of practical significance. The geology of north Lancashire and that of Derbyshire are closely similar, with a preponderance of lower Carboniferous limestones and, presumably in both areas, with a deficiency of iodine in the drinking water. Yet simple iodine-deficiency goitre is much commoner in Derbyshire ('Derbyshire neck') than in north Lancashire, and this is accompanied by an appreciably higher frequency of tasters in the low-goitre area. On the above selective hypothesis we should expect that the excess of goitre-susceptible non-tasters in Derbyshire would (in the absence of goitre prophylaxis) be gradually reduced, and with it the incidence of goitre would also be lowered.

Linkage equilibration

The causes of change in gene frequencies are usually listed as mutation, genetic drift, and natural selection. But there is another process, to which attention has only recently been drawn, which can cause changes in the frequencies, not of single genes, but of haplotypes in closely linked systems.

We have repeatedly referred to linkage equilibrium in systems such as Rh, MNSs, Kell, and histocompatibility. Each of these involves three or more closely linked loci. If there is no natural selection favouring particular haplotypes in a given system, then in the course of time crossing over will distribute the alleles at one locus so that the ratios between the frequencies of combinations of the genes at a second closely-linked locus will be the same as the ratios between the total frequencies of these genes in the population under consideration. Thus if the total frequencies of the genes at one locus are 50, 30, and 20 per cent, then the combinations with a closely linked gene at another locus, with a total frequency of 10 per cent, should be 5, 3, and 2 per cent respectively.

It is inevitable that this process of equilibration should go on, but it may in part be counteracted by natural selection under a given set of circumstances favouring a particular haplotype combination.

There are two main possible causes for a lack of linkage equilibrium. Either selection is favouring those haplotypes found in excess of equilibrium levels, or the population is the result of recent mixing of two (or more) separate populations. Thus frequencies near equilibrium suggest that the population has existed in a condition of relative isolation for a period probably of several thousand years, while lack of equilibrium suggests one of the two alternatives just mentioned. Where such a lack of equilibrium is found, some authorities favour natural selection as a cause while others favour recent hybridization.

Thus, apart from history and archaeology, there are two independent genetic indications that the population of Europe is the result of relatively recent hybridization. One is the presence at relatively high frequencies of both of the Rh alleles *D* and *d*, one of which should, in an equilibrated population, have been eliminated by haemolytic disease of the newborn (pp. 74–5). The other is the high general degree of linkage disequilibrium for the histocompatibility system, to which Dr L. Degos has drawn attention and which he, unlike others, attributes to hybridization.

The high frequency of the *CDE* haplotype in Amerindian and Australian Aborigine populations appears to be due to crossing over between the haplotypes *CDe* and *cDE*, and is consistent with each of these great groups of populations having existed in isolation from the outside world for many thousands of years.

Conclusion

Genetic anthropology has two complementary aspects. In so far as gene frequencies are stable from generation to generation, the genes in a population provide a very precise index of the biological relationship of one population to another. But such information can be used only if populations differ genetically, and the differences between them can have arisen only through an instability of the gene frequencies.

The study, described earlier in the present chapter, of the mechanisms of change in gene frequencies from generation to generation in a single population, and of the rates of such changes, helps us to interpret differences between populations in a more or less quantitative manner. Some of the processes of change—mutation, genetic drift, linkage equilibration—are almost totally uninfluenced by the environment, but one of them, natural selection, is so dependent. Not only is the study of natural selection of blood groups and other genetic characters therefore important for the anthropological interpretation of gene frequencies, but it has a very important future in medicine, helping us to predict and thus to guard against genetic tendencies to disease in individual patients. As such tendencies become better and better understood it may be found that a very large part of disease susceptibility is hereditary, but subject to control by suitable modification of the environment in the widest sense, by diet, immunotherapy, and countless other means.

Further reading

BAKER P. T. and WEINER, J. S. (ed.) (1966). *The biology of human adaptability*. Clarendon Press, Oxford.

BODMER, W. F. and CAVALLI-SFORZA, L. L. (1976). *Genetics, evolution, and Man*. W. H. Freeman & Co., San Francisco.

BOYD, W. C. (1950). *Genetics and the races of man*. Little Brown and Co., Boston, Mass.: Blackwell Scientific Publications, Oxford.

COON, C. S. (1972). *The living races of man*. Jonathan Cape, London. Penguin, Harmondsworth (1976).

GIBLETT, E. R. (1969). *Genetic markers in human blood*. Blackwell Scientific Publications, Oxford.

HARRIS, H. (1979). *The principles of human biochemical genetics*. North-Holland, Amsterdam.

HARRISON, G. A. (ed.) (1977). *Population structure and human variation*. Cambridge University Press.

HOWELLS, W. W. (1960). *Mankind in the making: the story of human evolution*. Secker and Warburg, London. Penguin, Harmondsworth (1967).

HULSE, F. S. (1971). *The human species* (2nd edn). Random House, New York.

KOPEĆ, A. C. (1970). *The distribution of the blood groups in the United Kingdom*. Oxford University Press.

LEHMANN, H. and HUNTSMAN, R. G. (1974). *Man's haemoglobins*. North-Holland, Amsterdam.

MOURANT, A. E., KOPEĆ, A. C., and DOMANIEWSKA-SOBCZAK, K. (1976). *The distribution of the human blood groups* (2nd edn). Oxford University Press.

MOURANT, A. E., KOPEĆ, A. C., and DOMANIEWSKA-SOBCZAK, K. (1978). *Blood groups and diseases*. Oxford University Press.

MOURANT, A. E., KOPEĆ, A. C., and DOMANIEWSKA-SOBCZAK, K. (1978). *The genetics of the Jews*. Clarendon Press, Oxford.

PROKOP, O. and UHLENBRUCK, G. (1969). *Human blood and serum groups* (Trans. J. L. Raven). Maclaren and Sons, London.

RACE, R. R. and SANGER, R. (1975). *Blood groups in man* (6th edn). Blackwell Scientific Publications, Oxford.

ROBERTS, D. F. and SUNDERLAND, E. (ed.) (1973). *Genetic variation in Britain*. Taylor and Francis, London.

ROBERTS, J. A. FRASER and PEMBREY, M. E. (1985). *An introduction to medical genetics* (8th edn). Oxford University Press.

STEINBERG, A. G. and COOK, C. E. (1981). *The distribution of the human immunoglobulin allotypes*. Oxford University Press.

STERN, C. (1973). *Principles of human genetics* (3rd edn). W. H. Freeman & Co., San Francisco.

STRICKBERGER, M. W. (1976). *Genetics*. Macmillan, London.

TILLS, D., KOPEĆ, A. C., and TILLS, R. E. (1983). *The distribution of the human blood groups: Supplement 1*. Oxford University Press.

VOGEL, F. and MOTULSKY, A. G. (1979). *Human genetics: problems and approaches*. Springer-Verlag, Berlin.

WEINER, J. S. and HUIZINGA, J. (ed.) (1972). *The assessment of population affinities in man.* Clarendon Press, Oxford.

WEINER, J. S. and LOURIE, J. A. (ed.) (1969). *Human biology: a guide to field methods.* Blackwell Scientific Publications, Oxford.

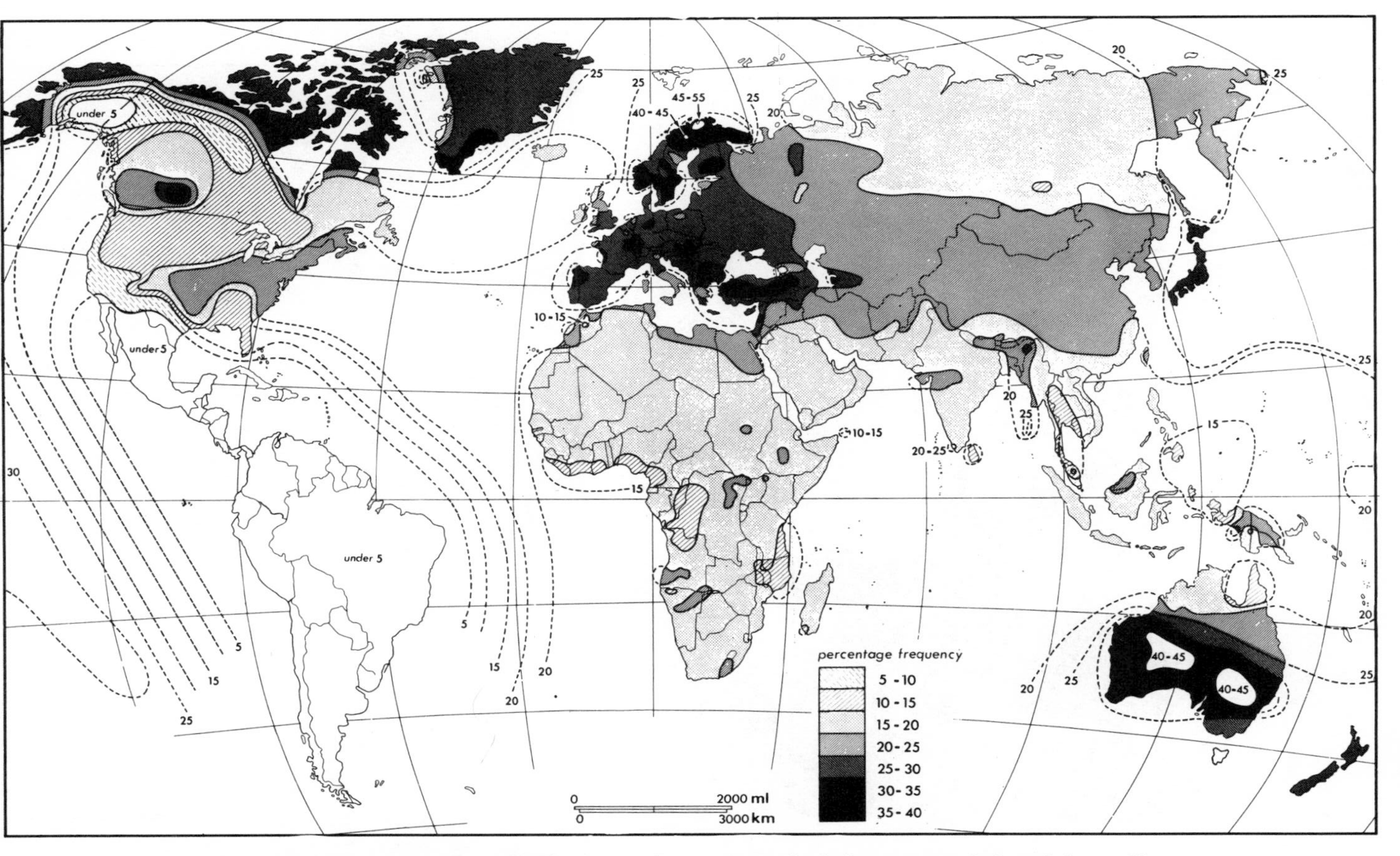

I. ABO blood group system. Distribution of the *A* gene in the indigenous populationsof the world.

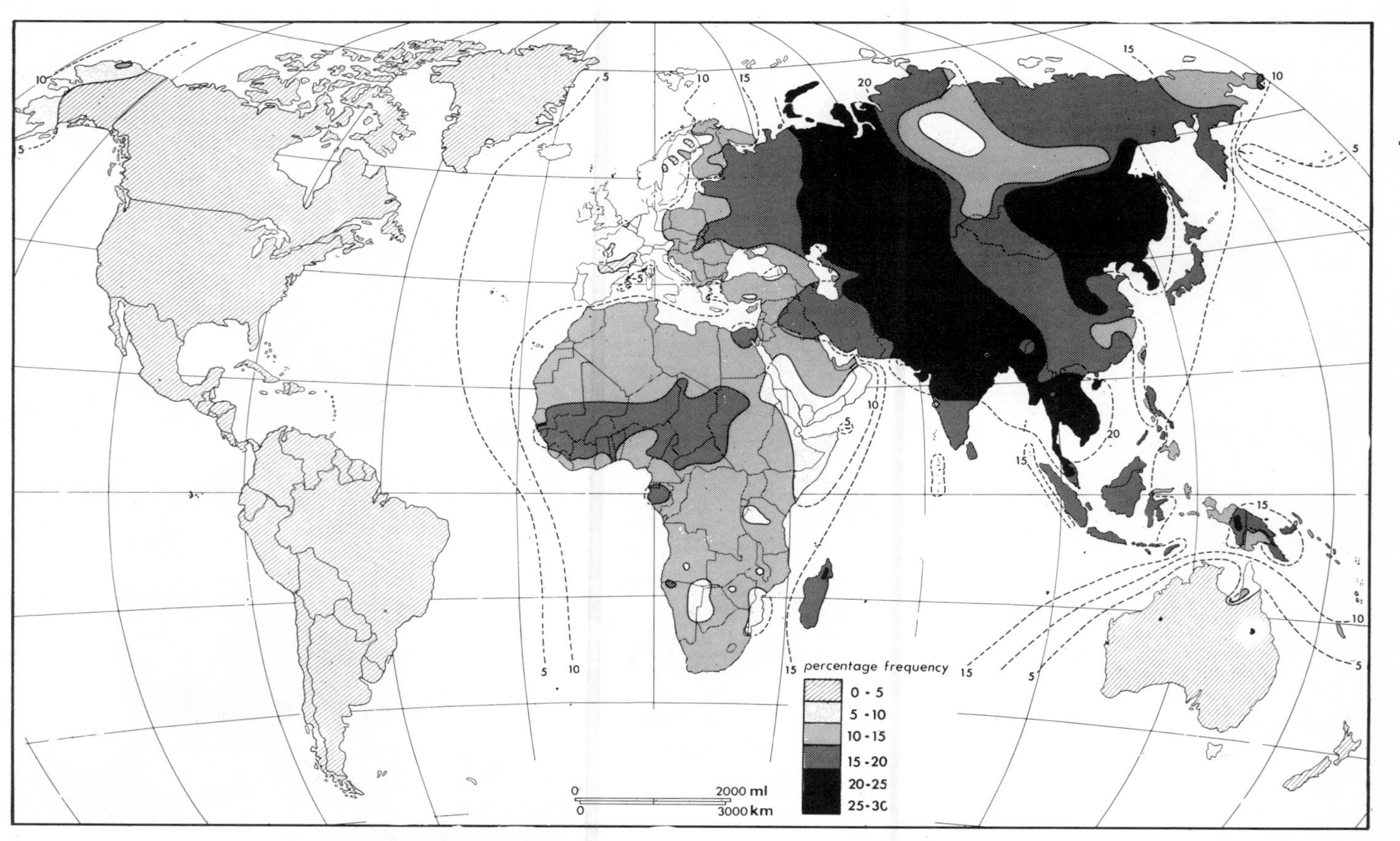

2. ABO blood group system. Distribution of the *B* gene in the indigenous populations of the world.

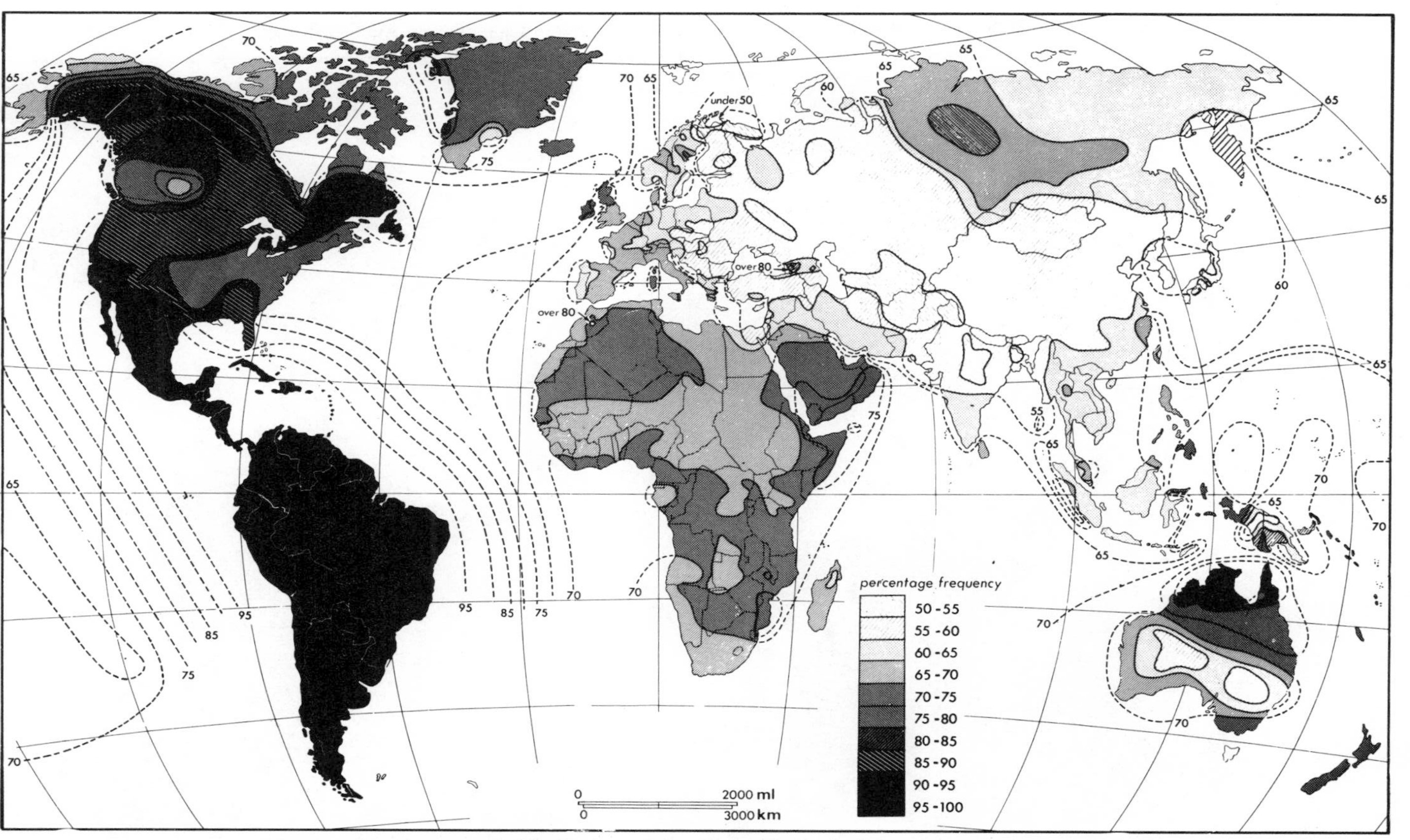

3. ABO blood group system. Distribution of the O gene in the indigenous populations of the world.

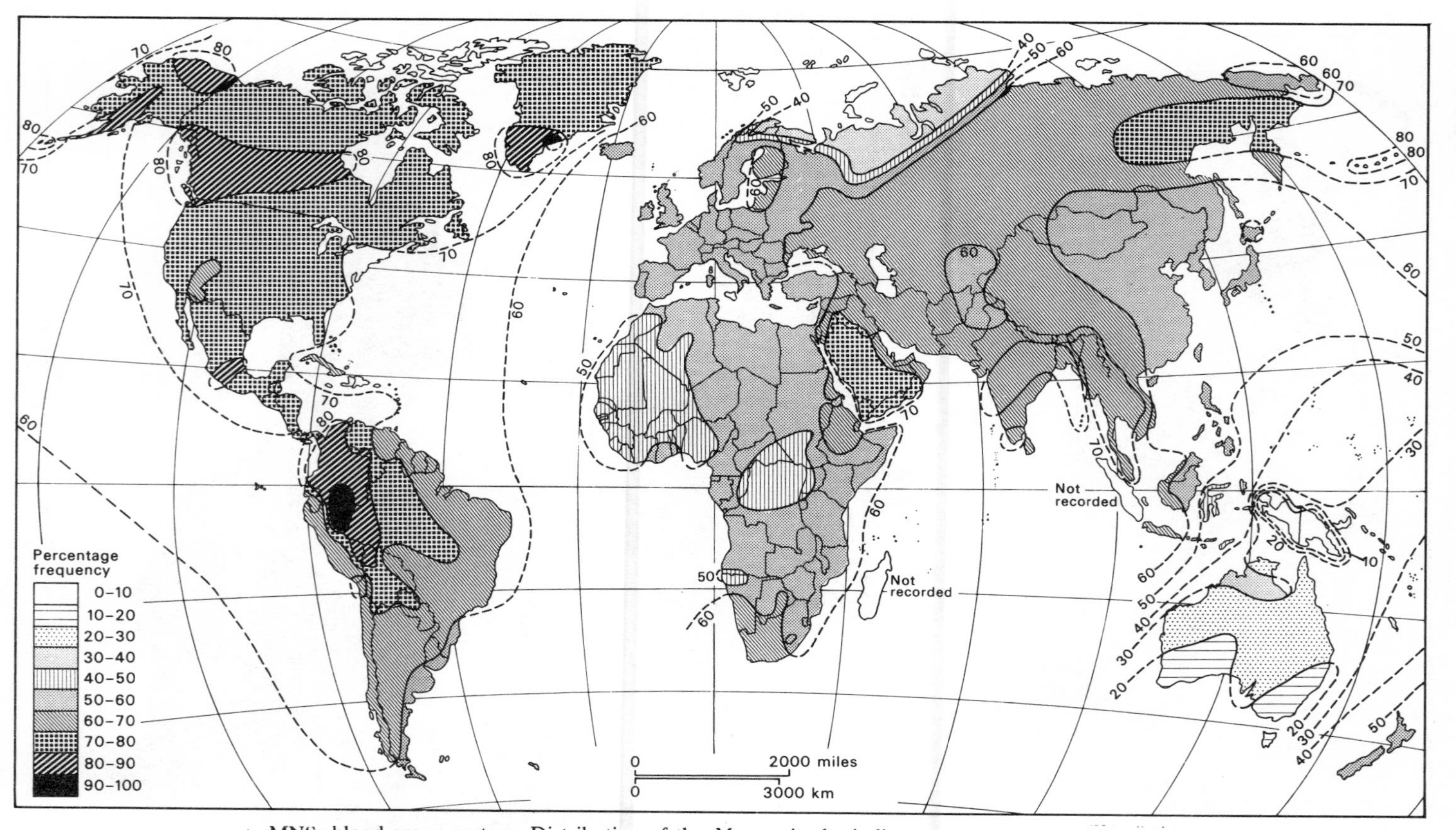

4 MNSs blood group system. Distribution of the *M* gene in the indigenous populations of the world.

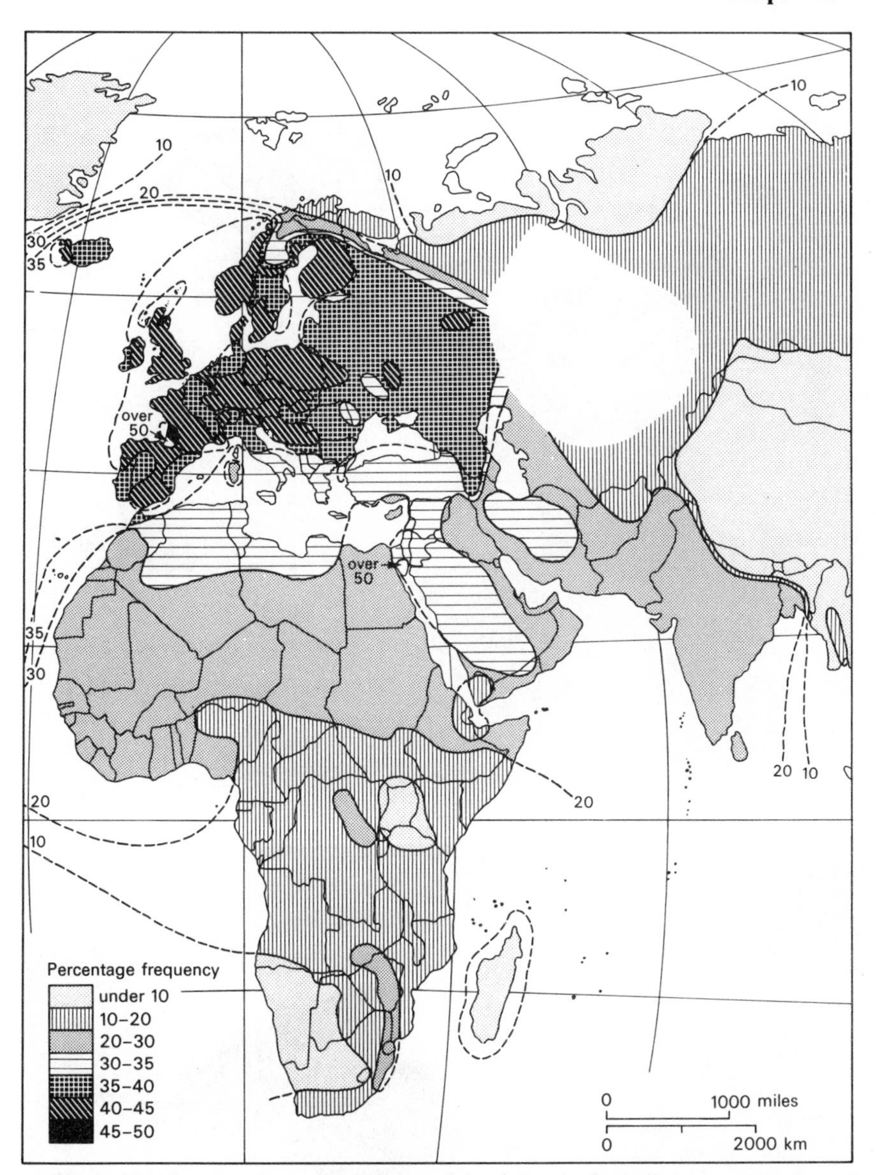

5. Rhesus blood group system. Distribution of the *d* gene in the indigenous populations of Europe, Western Asia, and Africa.

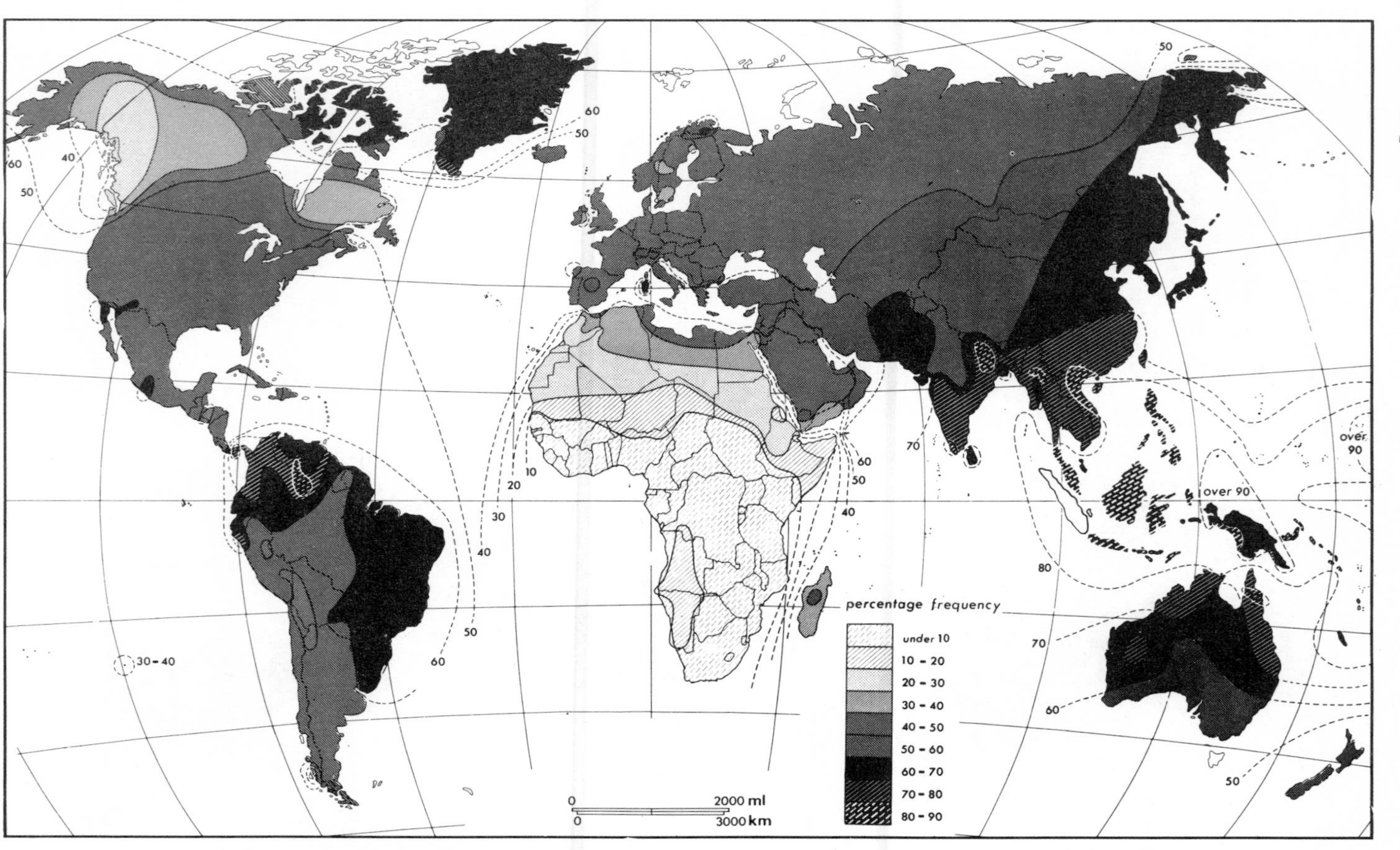

6. Rhesus blood group system. Distribution of the *C* gene in the indigenous populations of the world.

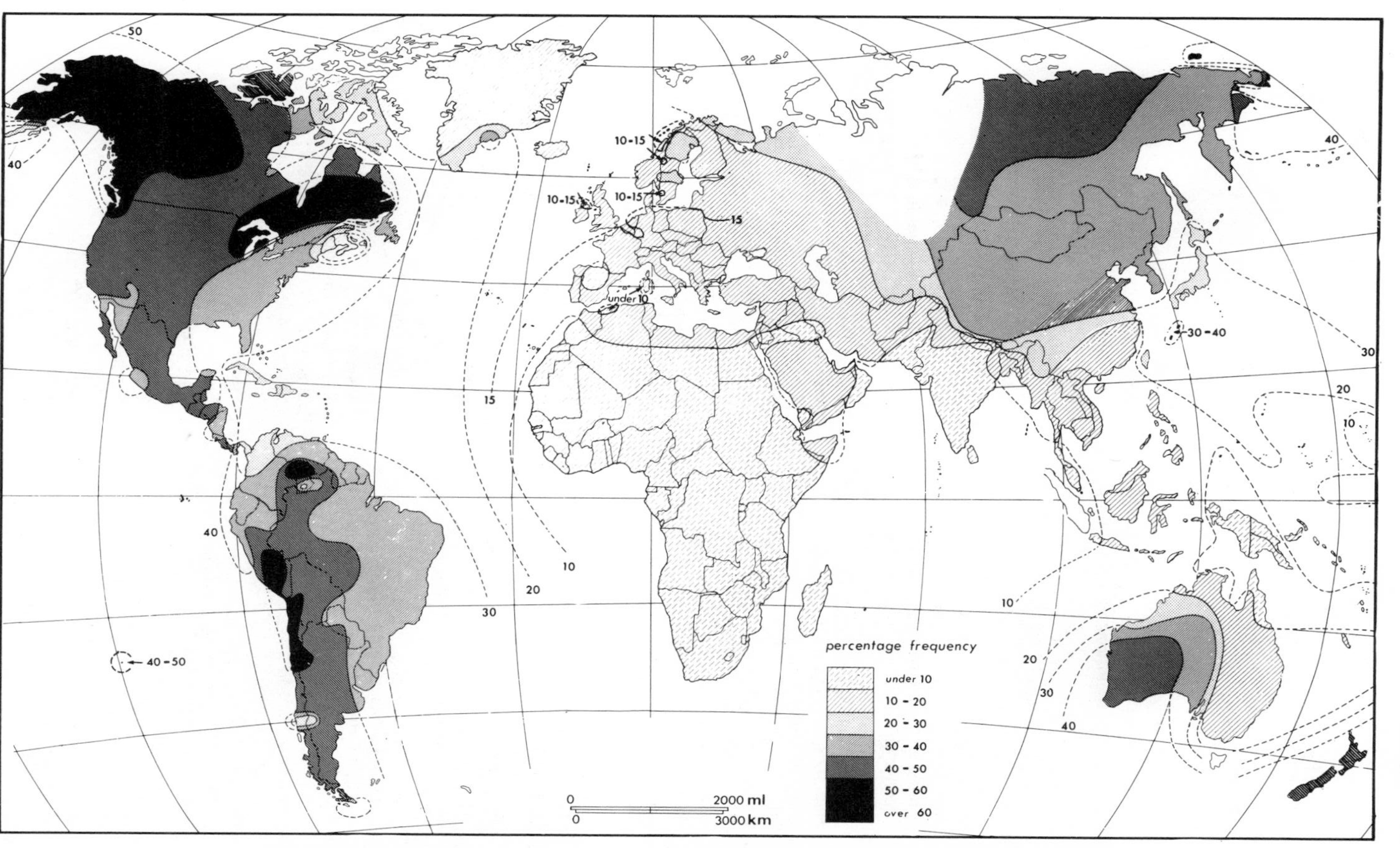

7. Rhesus blood group system. Distribution of the *E* gene in the indigenous populations of the world.

Index

Pages where definitions of technical terms occur appear in **bold type**

A gene
  high frequencies 82, 106
  map 133
A_2 gene 53, 78
Abd Kulalem 95
ABO blood groups 9, 10
  antigens and antibodies 10
  genotypes 10
Abraham 89
Abramsky, C. 94
acid phosphatase system **31**, 39
acquired characteristics 1
adenine 23
Adriatic Sea 82
Afars and Issas, Territory of 56–7
Afghanistan 71
Africa 30, 36–58
  east 37, 43–6
  north 37
  north-east 56–7
  west 58
African admixture in Jews 97
African Jews 96
African languages 44
African slaves 70
Africans
  genetic variability 40
  physical appearance 37
  west 41
Afro-Asiatic languages 44, 49, 56
agglutination **9**, 10
Ahaggar 55
Ainu 62, 63, 72, 78
Aïr highlands 56
Ait Haddidu 53–4
Alaska 112
Alaska Peninsula 112
Aleutian Islands 112
Aleuts 112
Alexandria 91
Algeria 53
Algiers 53
Allison, A.C. 126
America 63
Amerindians 64, 107, 108–14
  middle 110–11
  north 110
  south 111–12
Anatolia 71
ancestors of man 36
Andaman Islands 61
Andes, *A* and *B* genes in 109
Andriana 47
Anglo-Saxons 81
animal blood groups 10–11
ankylosing spondylitis 127
Anthropologie 13
anthropology v
antibody **10**
antigens **10**
  comparison of blood groups and
    bacteria 117
apes, great 36–7
Arab invasions of Africa 49
Arabic language 50
Arabs 50, 52–3, 58, 69–70
Aramaic 92
Armenians 71
arteriosclerosis 125
Aryan languages 66
Aryans 67
Ashkenazim 93
Asia 59–72
Asia minor 71
Assam 68
Assyria 90
Athabascan Indians 113
Australia 59, 99
Australian aborigines 61, 100, 101, 130
Australoids 59, 67, 99–100
Austria 82
Austronesian languages 102
auto-immune diseases 118–19, 127–8

B gene, high frequencies 82, 83
B gene map 134
Babylon 91
Babylonia 90
Babylonian Jews 94
bacteria 9, 116, 117

balanced polymorphism **118**, 128–9
Bali 104
Balkan Peninsula 71
Baltic Sea 82
Baluchistan 66
Bantu speakers 41–3, 45
  and Khoisan 42–3
Banu Hilal 50, 54
Barrow, Point 112
Basques 74–6, 82
Bathgamuwa 60
Beja 56–7
Beni Ounif 55
Berbers 50, 58, 92
Bergamo 87
Bering Sea Mongoloids 113
Bering Strait 63, 108
Bernstein, F. 11
Bhutan 64–5, 68
Bielsa Valley 85
Black Death 93
blood groups v, 3
boats 108
Bodmer, W. 73
body characters 34–5
body odour 2
Bothnia, Gulf 82
Brahui 66
Brescia 87
Bretteville 81
Britain 79
Brittany 81
Bulgaria 88
Burma 61, 100
Burmese 59

C gene map 138
Cambodia 61
Canada 112
cancer 118
  and auto immunity 119
  and group A 119
Carcassi, V. 86
Caribs 112
Carpathian Mts. 87
Carpentaria, Gulf 100
Caspian Sea 71, 83
Caucasoids 59
  of Asia 65
Caucasus mountains 74
Cavalli-Sforza L.L. 73
CDE haplotype 101, 130
cDE haplotype 106, 111
cells 4–5
Caspian culture 49
cereal cultivation 40
Chaambas 55
Chad 57–8
Chaldeans 91
Charlemagne 93
Chenchu 60
Cheremiss 83
China 65
Chinese 60, 64
Chleuh 56
Chromatids **25**
chromosomes
  biochemistry of 22–3
  banding of 21
  formation of 25
  human 22
  sex 26
  X and Y 26
Chukchi 64
click languages 46
clot 9
code, genetic 23
conclusion 130
Congo–Kordofanian languages 40
conversion to Judaism 90, 93
Cook, Capt. W. 108
Cook Islands 106
Copais, Lake 88
Copts 51–2
Cornwall 81
correlations, false 117
Correns, C. 8
Corsica 85–6
cotton 105
Crick, F. 23
Crimea 93, 94, 97
Crookston, M. vi
crossing-over **12**, 25–6, 130
crusades 93
$C^{w}De$ haplotype 78, 82
Cyrus 91
cytology 5
cytosine 23
cytoplasm 5, 23
Czechoslovakia 82, 83

d gene map 137
Dahiyah-al-Kahina 92
Dalmatia 93
Damascus 90
Darlington, C.D. 6
Darwin, C. 4, 5, 6
Dasas 66
death, cause of 22
Derbyshire neck 129
desert populations 121
desoxyribonucleic acid **23**
Devon 81
De Vries, H. 8
Diaspora 90
diploid **5**, 22

Diego blood groups 30
Dinka 46
disease susceptibility, hereditary 130
Djerba 52–3
DNA **24**
Dodecanese 88
dog blood groups 11
Dog River 70
Domaniewska-Sobczak, K. v
dominant 7
Dravidian 66–8
drift, genetic, *see* genetic drift
Drosophila 8, 12
Duffy blood group system 30
  and malaria 39, 121–2
duplication of chain 23
duodenal ulcers 125

E gene map 139
Easter Island 105, 106
egg cell 4
Egypt 51–2
Egyptians 54, 96
Ehrlich, P. 9, 10
Elbe, river 82
England 81
environment 117
enzymes **23**
Ethiopia 56–7
Eritrea 57
erythrocytes, *see* red cells
Eskimos 108, 112–14
Estonians 83
Ethiopia 93
Europe 73–88
  southern 84–8
European genetic uniformity 77
Exodus 90
eyes, colour 1
Ezra 91

face shape 1
Falasha 93, 96
Ferrara 87
fertility, differential 123–4
Fezzan 55
Finno-Ugric languages 71, 73, 83
Finns 77, 83
Fisher, R.A. 28
Flittas 53, 80
founder effect **18**
France 81, 93, 94
frequencies 14–17

gamete **5**
Ganges basin 68
Gc plasma protein system 30–1, 123
gene frequencies 14–16
  calculation of 16
  change in 17, 115–130
genes **6**
genetic code **23**
genetic drift **18**, 79, 115–16
genetics 4–35
Gerbich blood groups 30, 103
Germanic languages 82
Germany 94
glands, skin 2
glucose-6-phosphate dehydrogenase
    system 31, 39, 48, 86, 95
Greece 88
Greenland 112
groups, blood, *see* blood groups
guanine 23
Gypsies 98

Habbanite Jews 96
Hadza 46
haemoglobins 32, 125–6
  A 32
  C 126
  D 32, 126
  E 32, 61, 62, 126
  S 32, 46–8, 60, 87, 118, 125–6
haemolytic disease of newborn 28,
  74–5, 119–21, 124–5
hair colour 1
Halath 90
Haldane, J.B.S. 74
Hamada 56
Han 60
haploid **5**, 22
haplotype **29**
Harappa 66, 67
Haratines 56
Hardy–Weinberg equation **15**
Harper, A.B. 113
Harris, Harry 128
Hathor 90
Hawaii 105
helix, double 23, 24
Henshaw blood group antigen 29, 38
  linkage relations 41
Herodian dynasty 91
heterozygote **7**
Heyerdahl, T. 105
Himalayas 64–5
Hirszfeld, L. 11, 13, 19
histone **23**
histocompatibility system 3, 33, 127–8
  associations 33, 128
HLA, *see* histocompatibility
Hoggar, *see* Ahaggar
Hokkaido 62
Homo erectus 59, 61, 99
Homo sapiens 37, 59

homozygote **7**
Hova 47
Htalu 62
Hudson Bay 112

Iberian peninsula 50, 85
ice age 63
Iceland 79
India 66–8
Indian 59
Indian origin of Gypsies 98
individuals 2
Indo-European languages 66, 68, 70, 71, 73
Indonesia 40, 99, 103–4
Indonesians 61
inheritance mechanisms 4
inherited characteristics 1
inheritance of bloodgroups 11
Inquisition 92
invisible characteristics 3
iodine 34, 128–9
Iran 70–1
Iraq 70, 94
Irish 79
iron smelting 40
island populations 121
island-hopping 99
Itil 93

Jahweh 91
Japanese 62–4
Java 61
Jebeliyah 69–70
Jenkins, T. 42–3
Jericho 65
Jerusalem 91
Jews 49, 90–8
Jordanian Arabs 70
Judeo-Persian dialect 92
Justinian 69

Kabyles 53
Kalmus, H. 128
Karaites 94, 97–8
karyotype **21**
Kell blood groups **29**
Khalkidhiki **89**
Khazars 93–4, 97
Khmers 61
Khoisan 42–3
Kitchin, F. D. 128
Koestler, A. 94
Kodiak Island 112
Komandorskiye Islands 112
Komi 83
Kopeć, A. C. v 81
Koreans 62

Krimchaks 97
Kurdish Jews 95
Kurds 71, 95
Kurumbas 60

Ladino 92
Lancet 13
Landsteiner, K. 9, 20
Lapps 53, 72, 78
Lathyrus odoratus 12
Latvians 83
Lebanon 70
Lemba 96
Levine, P. 20
leucocytes, *see* white cells
Libya 52–3, 96
linkage 12
  equilibrium 18–19, 101, 106
  sex-linkage, X-linkage 27
Lithuanians 83
locus, loci 12
Lombok 104
lost tribes 91
Low Countries 82

M gene
  high frequency 112
  map 136
Madagascar 47–8
Maghrib 44, 50, 52–5
Magyars 77
Malagasy *or* Malgache 47–8
malaria, *see* Duffy blood groups; haemoglobin S; thalassaemia
malate dehydrogenase system 32
Malay peninsula 61, 99
Malayo-Polynesian languages 47
Malaysia 100
Malaysian culture 47
man, single species 37
maps 44, 133–9
Mari 83
mechanisms of inheritance 4
Medes 90, 91
Mediterranean Sea 84
meiosis **25**
Melanesians 61, 102–4
Mendel, G. 4, 5, 6, 7, 8, 14
Mesopotamia 66
Messenger RNA **24–5**
Metlili 55
Mexico 110
Micronesians 104–5
Milan 87
Misson, G. v
mitosis **22**, 25,
MN blood groups 20, 29, 121

Mohammed 69
Mohenjo-Daro 66
Mongoloid populations 30, 59, 60, 61
Mons 61
Morgan, T. H. 8
Morgenroth, J. 10
Morocco 53–5, 58
mountain populations 121
multiple sclerosis 119
mutation **17**, 115
myasthenia gravis 119

natural selection 18, 77, 116–29
Nazis 94
Near East 65
Nebuchadnezzar II 91
Negritos 61
Negro slaves 47, 51
Nehemiah 91
Neo-Babylonians 91
Neolithic revolution 65, 73
Nepal 64–5, 68
New Guinea 61, 99, 101, 103
  archaeology 102
New Zealand 105
Nicaragua 111
Niger 57–8
Nile valley 49, 66
Nilgiri Hills 59
Normandy 81
Normans 81
Northern Nilotes 46, 57
Nubians 52
Nuer 46
nucleus 5, 23

odour, body 2
O gene 11
  high frequencies 79, 80, 109
  map 135
Oran 53, 80
Origin of species 4, 6
ovule **4**
ovum **4**

P blood groups 20, 39
Pacific islanders 99–107
Pacific, migration across 105, 108
Padhu 60
Padua 87
Pakistan 71
Palestine 48, 68–9
Palestinian Arabs 70
Paniyans 60
Parsis 95
pea plant 6
Persia 93
Persians 71, 91
Pharaohs 70
phenylthiocarbamide, *see* taster system
Philippine Islanders 104
Phoenicians 49
phosphoglycerate kinase system 31
Piacenza 87
pig 105
plasma 9
plasma protein systems **30–1**, 123–4
Plasmodium falciparum 46–8; *see also* haemoglobin S
Plasmodium knowlesii 122
Po valley 87
Poland 83, 94
polymorphism, balanced 118, 128–9
Polynesians 105–7
Pompey 91
populations 2
protease inhibitor system **31**, 123–4
proteins **23**
proto-Polynesian language 105
Punjab 66
Punnett, R. C. 12
pure line **6**
purine **23**
pygmies 46, 47, 61
Pyrenees 85
pyrimidine **23**

Race, R. R. 28
rain-forest crops 40
recessive **7**
recognition of people 1, 2
red-cell enzyme systems 31–2
red cells **9**
red corpuscles, *see* red cells
reduction division 25
redundancy **24**
Reguibat 56
reindeer herders 72
reproduction 4–5
reproductive cell **5**
Rhesus blood groups 27–9, 76, 121
rheumatoid arthritis 119
Rhineland 93
Rhodes 88
ribonucleic acid **24**
Rigveda 66
RNA **24**
Roberts, D. F. v
Romania 87–8
Rome 91
Ruffié, J. 55

Sahara Desert 55–6
St Catherine's monastery 69
Sakhalin 62
Salonika 13

Samaria 90, 91
Samoyeds 72, 78
Sandawe 46
Saoura 55, 56
Sardinia 84–6
Sargon II 90, 95
Saudi Arabia 69
savannah 41, 57–8
Scandinavia 79
Scots 79
Scythia 93
sea level, low 99
Second World War 70
secretor system 30, 112, 124–5
segregation **7**, 12
selection, natural, *see* natural selection
Semitic languages 68
Senegal 58
Senoi 62
Sephardim 92–3, 96–7
serum **9**
sex chromosomes **26**
sex-linkage 27
Shalmaneser V 90
Shilluk 46
Shipibo 111
Siberia 63–4, 112–14
Sicily 87
sickle-cell haemoglobin 32, 46–8, 60, 87, 118, 125–6
Sinai 69
Sinai Bedouin 69–70
skin colour 1
skin glands 2
skull shape 2
Slavonic languages 77
Somalia 56–7
somatic cell 5
Southampton Island 112
Soviet Asia 63, 72
Soviet Europe 72
Soviet Union 71
Spain 92
Spanish West Africa 56
spermatozoon **4**
spiral structure 23
Sri Lanka 60, 62, 66, 67, 100
SsS^{u} 38
statistics of disease 117
stature, human 7
stratification of populations 117
Sudan 56–7
Sudanic languages 44
Suez, Isthmus of 48
susceptibility to disease 117–30
Switzerland 82
Syria 70

Tafilalet casis 96
Taiwan 62, 104
Tarons 62
Tasmanians 102
taster system 33–4, 128–9
Temple destroyed 91
Thais 61
thalassaemias 32, 39, 87
thymine 23
Thirteenth tribe 94
thrombosis 118
thyroid
  disease 120
  hormones 34
  inhibitors 128
Tibet 59, 65
Tibetans 62
Tibeto-Burmans 61
Tidikelt 55
Titus 91
Titus, Arch of 91
Tizi Ouzou 53
tool-making 36
Touaregs 50, 54
Towara 69–70
transfer RNA **24**
Trieste 87
Trotter, W.R. 128
Tschermak, E. 8
Tunisia 52–3
Turkestan 71
Turkey 88
Turkish languages 68, 71, 74
Turkmen 71

Udmurts 83
ulcers, gastric and duodenal 125
Uzbekistan 71
Uzbeks 71

V antigen 38, 45, 62, 111
Valle Ladine 87
Veddas 60, 62
Veddoids 59–60
Venice 87
visible characteristics 3
Vitamin D 101
Von Dungern, E. 11
Votyaks 83

Wahabi 53
Wallace, A. R. 104
Wallace's line 104
Warao 112
Watson, J. D. 23
Weismann 5
white cells 9

white corpuscles, *see* white cells
Wiener, A. S. 28, 74

X chromosomes 26

Y chromosomes 26
yam 105
Yemenite Jews 69, 92, 95–6
Yugoslavia 83, 87, 88
Yukagir 64

Zambesi river 46–8
Zebu, short-horned 46–8
Zoutendyk, A. 42
zygote 5

GADSDEN
PUBLIC
LIBRARY